The Ethos of the Climate Event

This book develops an ethico-political response to climate change that accounts for the novelty and uncertainty that it entails.

This volume posits that an ethos can effectively connect the subjective, social, and environmental dimensions of climate change in order to understand and overcome this challenge. To show how this ethos builds upon the need for new forms of responsiveness, Anfinson analyzes it in terms of four features: commitment, worldly sensitivity, political disposition, and practice. Each of these features is developed by putting four thinkers—Kierkegaard, Nietzsche, Schmitt, and Foucault, respectively—in conversation with the literature on climate change. In doing so, this book shows how social habits and norms can be transformed through subjective thought and behavior in the context of a global environmental crisis.

Presenting a multidisciplinary engagement with the politics, philosophy, and science of climate change, this book will be of great interest to students and scholars of climate change, environmental politics, environmental philosophy, and environmental humanities.

Kellan Anfinson is Visiting Assistant Professor in the Department of Political Science, Bard College, USA.

Routledge Advances in Climate Change Research

For more information about this series, please visit: www.routledge.com/Routledge-Advances-in-Climate-Change-Research/book-series/RACCR

The Ethos of the Climate Event
Ethical Transformations and Political Subjectivities

Kellan Anfinson

Routledge
Taylor & Francis Group

LONDON AND NEW YORK

from Routledge

First published 2021
by Routledge
2 Park Square, Milton Park, Abingdon, Oxon OX14 4RN

and by Routledge
52 Vanderbilt Avenue, New York, NY 10017

Routledge is an imprint of the Taylor & Francis Group, an informa business

British Library Cataloguing-in-Publication Data
A catalogue record for this book is available from the British Library

Library of Congress Cataloging-in-Publication Data
Names: Anfinson, Kellan, author.
Title: The ethos of the climate event : ethical transformations and political
    subjectivities / Kellan Anfinson.
Description: Abingdon, Oxon ; New York, NY : Routledge, 2021. |
    Series: Routledge advances in climate change research | Includes
    bibliographical references and index.
Identifiers: LCCN 2020040041 (print) | LCCN 2020040042 (ebook) |
    ISBN 9780367484576 (hbk) | ISBN 9781003044192 (ebk)
Subjects: LCSH: Climate change mitigation—Moral and ethical aspects. |
    Climatic changes—Philosophy. | Climatic changes—Political aspects. |
    Environmental justice.
Classification: LCC TD171.75 .A45 2021 (print) | LCC TD171.75 (ebook) |
    DDC 304.2/5—dc23
LC record available at https://lccn.loc.gov/2020040041
LC ebook record available at https://lccn.loc.gov/2020040042

ISBN: 978-0-367-48457-6 (hbk)
ISBN: 978-1-003-04419-2 (ebk)

Typeset in Times New Roman
by Apex CoVantage, LLC

Contents

And we say that the world isn't dying.
And we pray that the world isn't dying.
And just maybe the world isn't dying.
Maybe she's heavy with child.
    —Jason Webley, Last Song

Acknowledgments

Thanks goes to colleagues and comrades beyond those listed here. Derek Denman, Chitra Venkataramani, Tim Hanafin, and Filip Wojciechowski provided the intellectual ferment that cohabitation enables. Derek and Chitra read many drafts and gave bountiful feedback. My mentor William E. Connolly has been a constant source of intellectual guidance. Jane Bennett has been generous with her patience and insight. Simon Gilhooley, Pınar Kemerli, Dorothy Kwek, Olivier Ruchet, and Robert Seyfert helped more than they know. Kiarina Kordela, David Blaney, and Franklin Adler gave me roots. The Max Kade Center at Johns Hopkins and the Hong Kierkegaard Library provided material support. The feedback I've received from interlocutors and reviewers has been crucial. Telos Press and Taylor and Francis generously gave permission to reprint parts of Chapters 4 and 5. Last but not least, I am grateful to all the students who have explored the climate event with me. It is sometimes unclear to me whether I have contributed anything more to this project than being a medium through which these and other influences were able to interact. In this sense, it also seems appropriate to thank contingency.

Introduction

> *As a man divinely abstracted and self-absorbed into whose ears the bell has just drummed the twelve strokes of noon will suddenly awake with a start and ask himself what hour has actually struck, we sometimes rub our ears after the event and ask ourselves, astonished and at a loss, "What have we really experienced"—or rather, "Who are we, really?" And we recount the twelve tremulous strokes of our experience, our life, our being, but unfortunately count wrong.*

In these opening lines to the *Genealogy of Morals*, Nietzsche uses the event to awaken us to the fact that we do not understand ourselves, the purposes driving our existence, or where they are leading. For us, the bell of climate change has tolled. Unlike many theories that see the event as a dramatic unveiling, Nietzsche emphasizes how we might miss it or even lose ourselves within it. Indeed, the original German text does not reference the event directly, allowing it to slip by the inattentive reader unheeded, with the word "afterward" as the only sign of its occurrence. The goal, however, is not to nail down the fact of the event. Rather, tracing its reverberations can help us understand our relation to it and how our existence is constituted. Indeed, dwelling in the existential uncertainty of the event may be necessary if it is to be proactively shaped. Today, we are lost within the climate event, seeking ways to give form to the future.

When I began this project, there was little if any reference to climate change as an "event." In the last 5 years, it has become widespread.[1] Yet hardly any of these works discuss what it means if climate change is an "event." One way to do so would be to delve into the extensive literature on the event, which has been used to explore issues from explosive revolutions to the ephemeral knitting together of being in each moment that passes, and ask what that means in the context of climate change. This literature has focused primarily on defining the nature of the event and the structure of its emergence and functioning. Less attention has been given to the event as something that can be ignored, co-opted, repressed, devalued, sublimated, or covered up. As Nietzsche points out in the epigraph that opens this chapter, the relevance of an event does not mean that it is evident or recognized. The widespread incapacity to adequately confront climate change attests to this. Thus, the question that motivates this study: If an event such as climate change is diffuse, complex, and elusive, what kind of ethos is needed to recognize, engage, and respond to it?[2]

An event is an indefinite break or transition in established processes. By an ethos of the event, I mean both the sensibility we develop in a world understood to be punctuated by events and the existential-spiritual responses most appropriate when we encounter an event. There are two dimensions to such an ethos. The first is an affirmative disposition toward a world and life without permanence, transcendence, or the promise of inevitable improvement in the human condition; a world that periodically includes dramatic changes, unforeseen developments, and sometimes tragic turns of chance. The second is the spiritual character and capacities necessary to recognize events, engage the discomfort they entail, and transform established ways of living through them. Though I focus on the subject as a critical site for thinking through the question of ethos, I argue that the subject is not a self-sufficient individual but rather a social site whose thought and action are embedded in politics. This book argues for an ethos to effectively engage the ethical dimension of politics as it encounters the novelty of climate change. In particular, it looks for ways not only to encourage a willingness to dwell in uncertainty and discomfort but also to act in the face of those obstacles.

I develop this ethos, in terms of four features: commitment, worldly sensitivity, political disposition, and practice. Each of these features is developed by putting four thinkers—Kierkegaard, Nietzsche, Schmitt, and Foucault, respectively—in conversation with the literature on climate change. In doing so, I show how this ethos can transform social habits and norms through subjective thought and behavior in the context of a global environmental crisis. Bridging these multiple levels, this ethos helps us to work through the problematic collective anthropos presupposed by the Anthropocene, both in terms of locating the cultural drives of human behavior and in composing a collective subject that can respond to climate change. Ethos is thus not simply an intellectual matter, but a material one, from the worldliness of Kierkegaard to the practices of Foucault. The importance of this ethos is underscored by the complex milieu of media, consumption habits, scientific information, spiritual dispositions, and personal aspirations through which people relate to climate change.

The debate around the term "anthropocene" and who the subject of climate change is will be taken up in Chapter 3, but a short terminological discussion is necessary here.

I will use the pronoun "we" throughout this book, a term that is both problematic and important. At its heart, the problem is that no meaningful form of "we" that can accurately designate the people, regimes, inequalities, temporalities, histories, ideologies, or geographies that have contributed to or will be affected by climate change. To suggest otherwise is to begin erasing and justifying the inequalities and injustices built into climate change. Yet if the climate crisis cannot be understood or responded to without accounting for the divisions and power imbalances central to the evolution of the global carbon economy, those divisions are not solid enough to define a responsible agent for responding to it.[3] Beyond these divisions, there is no "we" with the capacity and efficacy to meaningfully act on climate change.

Because no "we" that can act responsibly currently exists, it is necessary to compose one. "Nature is . . . the assemblage of contradictory entities that have to be composed together. This work of assembly is especially necessary if we now are to imagine the 'we' that humans are supposed to feel part of in taking responsibility for the anthropocene."[4] While this must be done with attention to power imbalances and questions of justice, to focus on a collective subject is already an important move against the way individualizing systems of politics, economics, and ethics elide and perpetrate systemic injustice.[5] This entails work on the self to cease treating ourselves as individuals and promote engagements that build and enhance social collectivities, as well as efforts to enhance cross-species connections.[6]

Many people, liberal theorists and political subjects in particular, believe that a responsible "we" already exists and that they are part of it. This is the comfortable and unexamined position of many well-intentioned privileged people primarily in Anglo and European countries who blame climate denialists for the lack of action even as they perpetuate destructive modes of existence. To upset such presuppositions and engage the problem that no "we" exists, I use what I call the "auditorial we" in that it serves as an opportunity for the reader to audit their assumptions, concerns, ethics, politics, and aspirations in the face of climate change. The process of auditing is critical to the thinkers examined in this study, who find themselves in the middle of unfolding events and through this process find ways to connect to and help shape those events. In the process, each discovers not only that they are not singular individuals but internally plural tendencies, affects, and experiences but also that an external universal unification with others is neither possible nor preferable. The "we" to be composed is not a universal humanity of individuals but a differentiated multiplicity. As Latour points out, "we haven't finished absorbing the diversity of ways of occupying the Earth. The Anthropocene is first of all the opportunity to listen seriously at last to what anthropology teaches us about other ways of composing worlds."[7] The auditing process reveals limited borders and opens the way to build new connections to others through climate change. Thus, the "we" of this book is a "we" to come, a collective subject in the process of being constituted.

The climate event has many characteristics but four seem particularly salient: complexity, denial, the complication of agency, and violence. First, climate change is an uncertain, complex, and novel event. The science of the Earth system is so complex due to interplays and feedback that understanding the role of discrete elements such as melting ice or clouds still contains a lot of uncertainty. Add to this the highly probable but relatively weakly understood tipping points and the picture gets more complex.[8] Add to this the fact that greenhouse gas emissions are dependent on economies, politics, and other social and cultural institutions and mapping a likely future becomes more difficult still.[9] Scientific uncertainty, in turn, feeds back into political uncertainty which is considerable since no society has yet enacted policies in line with the best science available, which feeds back into scientific uncertainty. Beyond this, our understanding of climate change relies on the history of the climate, and greenhouse gas emissions are quickly rising beyond the point where we have data about that history, so that we may never have more certainty about climate change that we do now.[10] We are entering novel and unknown territory. These interwoven issues of complexity, uncertainty, and novelty make the climate event elusive in spite of its immensity and make responding to it difficult. This is not entirely negative, since a fourth term relevant to this complex is possibility.

Second is the issue of denial. Though climate change is "probably the best-established fact in the whole of natural history," it may also be the most widely denied through both direct and indirect means.[11] Overt religiously inspired denialism is well known, and the highly funded and well-organized efforts to promote doubt about climate change on behalf of economic interests have been well documented.[12] But a variety of more insidious forms of denial have arisen as well. At a basic level, this includes the problem of cognitive dissonance in which people likely to agree that climate change is a serious problem become skeptics through adherence to a competing commitment such as economic growth or human mastery over nature. Kari Marie Norgaard's *Living in Denial* goes further to look at how emotional denial prevents effective action among those who are rationally committed to minimizing climate change. William E. Connolly has located a kind of denial in what he calls "passive nihilism" or the "formal acceptance of the fact of rapid climate change accompanied by a residual, nagging sense that the world ought not to be organized so that capitalism is a destructive geologic force."[13] The effect is to magnify the gravity of everyday concerns and burdens and diminish commitment to stronger action on climate change. Perhaps most striking are claims that the very structure of science may be a significant source of climate skepticism through commitments to statistical stringency, treating the world discretely rather than as connected, and approaching the Earth as a passive background object of study.[14] The issue of denial is more significant than presumed and forms a serious obstacle to action, even among committed activists and climate scientists.

Third, the climate event is characterized by a new complication of agency. On one hand, it is marked by the emergence of a new threatening Earth-scale climatic agency, which Isabelle Stengers calls "the intrusion of Gaia."[15] On the other, this cannot be disconnected from the new human agency signified by the term Anthropocene, designating the human capacity to shape terrestrial processes that set climate change in motion. Neither humanity nor Gaia is singular but composed of multiple agents such as the ocean conveyer system, market economies, air conditioning, rainforests, and even carbon itself. Human and natural systems now interpenetrate, complicating any simple notion of agency. Classic agents, such as individuals, states, and organizations, no matter the resources and effort they put into minimizing climate change, are "able to do nothing about it. Nothing at the right scale."[16] It is unclear whether any existing notion or institutionalized version of responsibility can handle the temporal and division of labor dispersions of actions, knowledge, and effects that make up the climate event.[17] Thus, it is not surprising that Bruno Latour concludes that while no agent capable of responding to climate change currently exists, this means that the task is to compose one. Climate change is thus marked by the intensification, interconnection, and dispersion of agency.

Finally, climate change is an apparatus of violence and injustice. Climate change will directly cause the deaths of humans and other species. Most human death and suffering, however, will be linked to social systems of privilege and exclusion. Additionally, much climate violence will occur at a speed and scale outside of normal perception, meaning that it will likely not be recognized as violence.[18] Thus, the exposure of various populations to the environment and the ability of others not to have to think about environmental precarity are critical factors in how climate change will intensify existing systems of injustice.[19] Already, individuals with market advantages are seeking to consolidate those into survival and security advantages in the face of climate change.[20] States are pursuing an exclusionary "politics of the armed life boat."[21] Climate change is already deeply entangled in the history and effects of imperialism,[22] will fuel traditional forms of conflict, and lead to the creation of new forms of socially organized violence.[23] Not to mention the looming threat of extinction.

These four characteristics do not encompass the climate event but are critical components of it. They are critical because they both help define the shape of this event and make it difficult to definitively delimit. As is the nature of the event, it is characterized by that which exceeds our current capacities for characterization. This is how A. S. Byatt portrays climate change: "Nothing saw her coming, for she was too vast for their senses to measure or expect. She was the size of a chain of firepeaks: her face was as large as a forest of kelp, and draped with things that clung to her fronds, skin, bones, shells, lost hooks and threads of snapped lines. She was heavy, very heavy. She crawled across beds of coral, rosy, green and gold, crushing the creatures, leaving in her wake a surface blanched, chalky, ghostly."[24] Here is the emergence of a destructive and unforeseen agency whose scope is indeterminate and whose composition is a muddled mixture of nature and human artifice. Though not definitive, I would like to suggest that these four characteristics form an important touchstone for attempts to respond to climate change. I have only briefly outlined them here and will deal with them in more detail throughout the book.

Chapter 1 examines the dominant ethical response to climate change developed by analytic liberal theorists of justice. While a minority within that tradition suggests that climate change challenges its theoretical capacities, I argue that this tradition is caught in a restrictive mode of thought that makes it unable to think through the ethical and political implications of climate change as a novel and uncertain event. Indeed, such ethical approaches may even provide defensive justifications for the very institutions that are primary drivers of climate change. As an alternative to the liberal justice model, the chapter concludes by turning to the thought of Kierkegaard, who understands God's intervention into the world as an event that challenges universalizable ethics. This opens the way to a deeper exploration of ethical and political responses to the climate event in the following chapters.

Chapter 2 builds on Kierkegaard's response to the ethical challenge outlined in Chapter 1 to develop a passionate response to match the urgency of climate change. For him, because uncertainty defines the way the event challenges ethics, it is not something that can be solved but must be committed to through belief. Yet Donna Haraway's commitment to "staying with the trouble" of climate change includes a compelling case for rejecting belief. I argue that in pursuing belief, Kierkegaard opens up ways of feeling and engaging the trouble that productively build on Haraway's approach. Ultimately, for Kierkegaard belief is an intensification of passion. The chapter closes by suggesting that "impassioned trouble" as a form of commitment is a critical component of an ethos for the climate event.

Chapter 3 takes up the debate around the term "Anthropocene" and the correlate discussion of who is the subject of climate change. Against the sufficiency of any particular term, it argues for increased sensitivity to the interlinked levels of the human existence: subjectivity, social relations, and the environment. Nietzsche develops this sensitivity, calling our attention to how large and small events interrupt life and proposing an experimental ethos to shape them. One critical site where this ethos plays out is in how drives shape our experience. The significance of this dimension is discussed through Kari Marie Norgaard's analysis of the emotional denial of climate change, which she argues is socially and politically structured. The capacity to affirm unexpected emotional eruptions can help forge stronger social responses to climate change. The chapter concludes by discussing Nietzsche's take on the event of the human to affirm a world that is not necessarily predisposed to human survival, a particularly pressing issue in the face of extinction events.

Chapter 4 looks at how political institutions can be used to suppress the events that interrupt social life or can be transformed through them. I argue that the outcome is heavily influenced by the ethos that infuses political participation. While an ethos of security like Schmitt's favors decisions made by powerful leaders to secure the given order, an alternative ethos of risk is necessary to transform political institutions to justly accommodate those populations most threatened in emergency situations. The importance of this ethos is shown in current discussions of climate change. A number of states are likely to follow the Schmittean formula by seeking to contain the impacts of climate change through exclusive security regimes. Against this tendency, Wainwright and Mann critique the possible emergence of a new global sovereign to deal with the climate emergency. Alternatively, Bruno Latour has repurposed Schmitt's decisionism as a compositionist technique for building climate change responsiveness, and Catherine Keller has recovered an apophatic approach to the uncertainty and precarity of the climate event from Schmitt's limiting political formula. I argue that an ethos of risk is a critical component to supplement these responses for resisting potential fascist climate futures.

Drawing connections with the previous thinkers, Chapter 5 traces the transformations in Foucault's thought throughout his career, forming a rough but incomplete schema for how the ethos developed in this book fits together. This can be seen by looking at how Foucault built upon limit-experiences at the subjective level to understand society as transformed through events operating on multiple levels and temporalities in his genealogical method. His critical contribution, however, comes in his work on care of the self, where he takes up the question of ethos directly to ask how to become an athlete of the event. I argue that the ethos of caring for oneself is a critical way to forge positive responses to climate change. This ethos can be seen in the case of Paul Kingsnorth and The Dark Mountain Project he helped form to respond to climate change.

The final chapter draws on the previous chapters to assemble an ethos for engaging the climate event. It begins by exploring the potentials and limits of the scientific literature, emphasizing how to proceed amidst simultaneously increasing levels of knowledge and uncertainty. I then move to consider political-theological approaches to climate change to analyze sovereign political responses and alternatives informed by deep commitments of belief. Next, a cultural take identifies problematic cultural drives that push us deeper into the climate crisis and proposes ways to transform those drives into more responsible modes of living as a form of ethical engagement. Finally, I look at a tragic reading of climate change that enables us to love and participate in an unruly world that is sometimes hostile to human existence and to better understand our role in bringing about that hostility.

A tragic perspective is not necessarily bad. Uncertainty is not doom. The traps we have set in our economies, politics, philosophies, and cultures are not fail-safe. To shift to thinking about climate change as an event is to accept that answers cannot come from within the system that created the problem. Indeed, answers themselves are not the goal. The disparate group of thinkers brought together in this study sought ways of living that were less reliant on answers to provide existential security. By finding ways to transform themselves through the event, they were able to both shape those events and feel at home in a dynamic world. This is perhaps the greatest challenge that we face today: to feel at home in a world capable of climate change, to want such a world, and to want to live our lives in it. Let us become emigrants of the false security of business as usual and experimenters with new ways of living, partisans for justice, foragers of past worlds, and forgers of new ones.

Notes

1 Examples include Haraway's, *Staying with the Trouble*; Latour's, *Facing Gaia*; Ghosh's, *The Great Derangement*; Connolly's, *Facing the Planetary*; Keller's, *Political Theology of the Earth*; Rushkoff's, "Survival of the Richest"; Danowski and De Castro's, *The Ends of the World*; Stengers's, *In Catastrophic Times*; Bonneuil and Fressoz's, *The Shock of the Anthropocene*.
2 Melissa Lane highlights the importance of developing such an ethos in her field review "Political Theory on Climate Change."
3 Ghosh, *The Great Derangement*, chapter 2.
4 Latour, "Waiting for Gaia," 7.

 5 Ghosh, *The Great Derangement*, chapter 3.
 6 Dipesh Chakrabarty takes up the compositionist theme along these lines in "The Human Condition in the Anthropocene."
 7 Latour, *Facing Gaia*, 182.
 8 Pearce, *With Speed and Violence*.
 9 Hamilton, *Requiem for a Species*, chapter 1.
10 Edwards, *A Vast Machine*, 438–9.
11 Latour, "War and Peace," 53.
12 See for example Oreskes and Conway, *Merchants of Doubt*.
13 Connolly, *Facing the Planetary*, 9.
14 Oreskes and Conway, *The Collapse of Western Civilization*, 14–18; Latour, *Down to Earth*, 77.
15 Stengers, *In Catastrophic Times*.
16 Latour, "Facing Gaia," 109.
17 Welzer, *Climate Wars*, 16–18.
18 Nixon, *Slow Violence*.
19 Norgaard, "Climate Denial and the Construction of Innocence."
20 Rushkoff, "Survival of the Richest."
21 Parenti, *Tropic of Chaos*, 11.
22 Ghosh, *The Great Derangement*.
23 Welzer, *Climate Wars*.
24 Byatt, *Ragnarok*, 71–2.

Bibliography

Bonneuil, Christophe, and Jean-Baptiste Fressoz. *The Shock of the Anthropocene: The Earth, History and Us*. Translated by David Fernbach. New York: Verso, 2017.

Byatt, A. S. *Ragnarok: The End of the Gods*. Edinburgh: Canongate, 2012.

Chakrabarty, Dipesh. "The Human Condition in the Anthropocene." In *The Tanner Lectures in Human Values*. New Haven, February 18–19, 2015. https://tannerlectures.utah.edu/Chakrabarty%20manuscript.pdf.

Connolly, William E. *Facing the Planetary: Entangled Humanism and the Politics of Swarming*. Durham: Duke University Press, 2017.

Danowski, Déborah, and Eduardo Viveiros De Castro. *The Ends of the World*. Translated by Rodrigo Nunes. Malden, MA: Polity Press, 2017.

Edwards, Paul. *A Vast Machine: Computer Models, Climate Data, and the Politics of Global Warming*. Cambridge: The MIT Press, 2010.

Ghosh, Amitav. *The Great Derangement: Climate Change and the Unthinkable*. Chicago: University of Chicago Press, 2016.

Hamilton, Clive. *Requiem for a Species*. New York: Earthscan, 2010.

Haraway, Donna. *Staying with the Trouble: Making Kin in the Chthulucene*. Durham: Duke University Press, 2016.

Keller, Catherine. *Political Theology of the Earth: Our Planetary Emergency and the Struggle for a New Public*. New York: Columbia University Press, 2018.

Lane, Melissa. "Political Theory on Climate Change." *Annual Review of Political Science* 19, no. 1 (2016): 107–23. https://doi.org/10.1146/annurev-polisci-042114-015427.

Latour, Bruno. *Down to Earth: Politics in the New Climatic Regime*. Translated by Catherine Porter. Medford, MA: Polity Press, 2018.

Latour, Bruno. *Facing Gaia: Eight Lectures on the New Climatic Regime*. Translated by Catherine Porter. Medford, MA: Polity Press, 2017.

Latour, Bruno. "Facing Gaia: Six Lectures on the Political Theology of Nature." In *Gifford Lectures on Natural Religion*. Edinburgh, February 2013. www.bruno-latour.fr/node/486.

Latour, Bruno. "Waiting for Gaia. Composing the Common World Through Art and Politics." In *French Institute Lecture*. London, November 2011. www.bruno-latour.fr/sites/default/files/124-GAIA-LONDON-SPEAP_0.pdf.

Latour, Bruno. "War and Peace in the Age of Ecological Conflicts." In *Peter Wall Institute Lecture*. Vancouver, September 23, 2013. www.bruno-latour.fr/node/527.

Nietzsche, Friedrich. *The Birth of Tragedy and The Genealogy of Morals*. Translated by Francis Golffing. Garden City, NY: Doubleday, 1956.

Nixon, Rob. *Slow Violence and the Environmentalism of the Poor*. Cambridge: Harvard University Press, 2011.

Norgaard, Kari Marie. "Climate Denial and the Construction of Innocence: Reproducing Transnational Environmental Privilege in the Face of Climate Change." *Race, Gender & Class* 19, no. 1–2 (2012): 80–103.

Oreskes, Naomi, and Erik M. Conway. *The Collapse of Western Civilization*. New York: Columbia University Press, 2014.

Oreskes, Naomi, and Erik M. Conway. *Merchants of Doubt: How a Handful of Scientists Obscured the Truth on Issues from Tobacco Smoke to Global Warming*. New York: Bloomsbury, 2010.

Parenti, Christian. *Tropic of Chaos: Climate Change and the New Geography of Violence*. New York: Nation Books, 2011.

Pearce, Fred. *With Speed and Violence: Why Scientists Fear Tipping Points in Climate Change*. Boston: Beacon, 2007.

Rushkoff, Douglas. "Survival of the Richest." In *This Is Not a Drill*. London: Penguin, 2019.

Stengers, Isabelle. *In Catastrophic Times: Resisting the Coming Barbarism*. Translated by Andrew Goffey. Atlantic Highlands, NJ: Open Humanities Press, 2015.

Welzer, Harald. *Climate Wars: Why People Will Be Killed in the Twenty-First Century*. Translated by Patrick Camiller. Malden, MA: Polity Press, 2012.

1 The challenge to ethics

Climate change as event

There is a more or less implicit, tacit or presupposed image of thought which determines our goals when we try to think. For example, we suppose that thought possesses a good nature, and the thinker a good will (naturally to 'want' the true); we take as a model the process of recognition—in other words, a common sense or employment of all the faculties on a supposed same object; we designate error, nothing but error, as the enemy to be fought; and we suppose that the true concerns solutions—in other words, propositions capable of serving as answers. This is the classic image of thought, and as long as the critique has not been carried to the heart of that image it is difficult to conceive of thought as encompassing those problems which point beyond the propositional mode; or as involving encounters which escape all recognition; or as confronting its true enemies, which are quite different from thought; or as attaining that which tears thought from its natural torpor and notorious bad will, and forces us to think.

Gilles Deleuze, *Difference and Repetition*

An extensive literature from the analytical branch of philosophy has developed on political and ethical responses to climate change. In this chapter, I argue that this approach is caught in what Deleuze calls an "image of thought" that impedes its attempt to think through the political and ethical implications of climate change. Broadly, this image presumes: first, that humans are well-intentioned, reasonable, morally serious individuals dealing with a well-defined scientific object; second, that on the basis of this universalizable reason and goodwill, the task of philosophy is to formulate propositional ethical standards and policies for responding to climate change; and third, that the key tools for implementation are nation-states and global capitalist markets. These assumptions prevent analytic liberal justice theory from dealing with the critical characteristics of the climate event raised in the introduction such as uncertainty and denial.[1] While some analytic thinkers have begun to recognize these limitations, they find themselves unable to move beyond them, suggesting that the problem lies in the nature of the approach—its image of thought—rather than any particular application of it.

Despite the shortcomings of analytic liberal justice theory with regard to climate change, it is important to examine it. It is the most popular approach to climate change in political theory and philosophy. Analytic thinkers tend to be published by high profile university presses and are more likely to advise policy-makers and NGOs. Some even participate in authoring IPCC reports. I argue that because analytic theory presumes the efficacy of existing liberal institutions such as the nation-state and the global capitalist market, its conclusions justify them. This becomes a serious problem if institutions like capitalist markets cannot solve climate change but are in fact drivers of it. Then, such philosophies become ideological defenses against effective and just responses to climate change.[2] Ilija Trojanow's fictional glaciologist might say that such thinkers are "not living and breathing beings so much as middlemen of destruction."[3]

After examining the analytic approach, I will turn to Kierkegaard to suggest an alternative, which will serve as the orientation for the rest of the book. His thought helps us see climate change as an interruptive event characterized by uncertainty and novelty. For Kierkegaard, there is no preexisting rational approach or institutional framework that enables a universalizable response to the existence of God. Even to a Christian, God occurs as an event that calls existing ways of living into question. It is only through an uncertain engagement that transforms a person and compels them to create new ways of living that one can respond to such an event.

The limits of liberal justice

I begin with those liberal analytic views which treat climate change as an unexceptional problem that conforms to the modes of analysis that would be applied to any other problem. The essays collected in Henry Shue's *Climate Justice* are an example of a set of proposals that assume that the best responses conform to the liberal analytic framework and are made by existing institutional arrangements like the nation-state and capitalist markets. Even more, the book does not give any hint that climate change threatens these arrangements. A brief examination of the way Shue handles uncertainty, the kinds of responses that are possible, and the appropriate way to analyze them will show the limitations of his approach.

Shue's engagement with the problem of the uncertainty does not examine the issue of tipping points, and his analysis suggests severe but gradual impacts from climate change. At one point, he argues that there are some cases where the impacts are so severe, that "one can reasonably, and indeed ought to, ignore entirely questions of probability beyond a certain minimal level of likelihood."[4] While this appears to take uncertainty and severe impacts seriously, Shue then turns to limit and contain that possibility to fit within an analytical frame. Rather than take up the genuine unpredictability of climate change, he approaches it through probability. Once the qualitative dimension of uncertainty is eliminated, the nature of the impacts is further reduced in that nothing outside a budgetary approach to evaluating them is considered possible. Shue can then judge a limited range of acceptable responses with a cost–benefit analysis in which the costs of taking action must be deemed "non-excessive" in relation to the impacts (265).[5] Such an analysis can only be deemed legitimate on the assumption that the effects produce no system-altering impacts, i.e. that the individuals and political actors retain their relevant place within a structurally unaltered system of money and goods.

This point becomes clearer when one looks at the three responses that Shue considers possible: do nothing and let the crisis happen, restrict economic activity and standards of living, or develop green energy.[6] The first option he sees as irresponsible, the second as impossible, and the third as the only hope.[7] There are two problems here. First, this displays what Harald Welzer calls a "false alternative" in which "the "objective constraints" show their totalitarian logic."[8] His point is that solutions whose notion of possibility presumes elements that are part of the problem such as capitalist markets, technological solutions, and national borders also treat the death and suffering of others as acceptable. As we will see, the analytic justice model is plagued by such false alternatives. Second, Shue evinces a failure to think about possibility not just in the imaginative sense that one hopes for when confronted with such crises but also in the empirical sense. Real possibilities such as climate change resulting in new levels of militarized nation-state violence, collapse of the global economy, and extinction events driving biosphere collapse are not thinkable within Shue's parameters. Indeed, he seems unable to think beyond assumptions like human life being independent of other life and the perpetuity of the nation-state system and global markets. For Shue, the latter are determinative of the quality of human life and an objective constraint on politics. Thus, the best and most forward-thinking response which he can imagine is phasing out carbon through increased pricing paired with making alternative energy affordable.[9] But it is precisely because of assumptions like Shue's that Naomi Klein argues that in some ways, climate denialists understand the problem better than liberal climate reform advocates: any just response must dramatically reconfigure the economy.[10]

This is disappointing given that Shue's analysis could be improved if he looked beyond the methodological assumptions which restrict the world in the name of a myopic ethical vision. The popularity, force, and appeal of Naomi Klein's *This Changes Everything* suggest persuasive transformations outside of capitalist economic configurations. Similarly, even if it fell far short of doing what it needed to do, the Paris Agreement makes it clear that international law beyond the nation-state will be necessary to respond to climate change, a point made only clearer by Trump's withdrawal. Or from the other side, it takes only a passing familiarity with Greta Thunberg and the school strike movement to see that much more ethical force resides at the level of protest than in responses whose assumptions require a system of governance and economics to be upheld which, as Shue notes, has failed to do anything thus far. None of these alternatives are solutions on their own, but they are empirical realities that suggest ethical and political responses outside of what Shue considers possible. Climate change, which was unthinkable as a possibility and the possibilities of which are not yet known, suggests the need to follow the advice of Lenin and try to be as radical as reality itself if we are going to find a solution for it.

Steve Vanderheiden's *Atmospheric Justice* also approaches climate change from within the analytic justice image of thought. For Vanderheiden, there is no novelty in climate change and he does not address the uncertainties of tipping points, feedback, and abrupt climate change. The only uncertainty that he does address is that promoted by the climate denial industry. While this is an important component, it is insufficient. As with Shue, this approach to uncertainty fails to address some of the more troubling potentials of climate change in favor of a reductive notion of scientific knowledge, reason, and the place of humans in the world.[11]

Yet even the problem of the climate denial industry proves to be insuperable for Vanderheiden's analysis. He argues that "we cannot adequately address global environmental problems that we do not fully understand," which we cannot do because of the climate denial industry.[12] At the same time, an effective response requires "the sincere search for truth and for solutions to common problems."[13] Here his theoretical assumption about the good nature of humans as rational individuals has put him in a bind that makes it impossible for him to engage the reality of the situation. For him, a responsive regime simply "must find a way" to be more democratic and exclude the influence of the denial industry.[14] Thus, Vanderheiden assumes magical mechanisms in order to preserve the image of thought. At heart is his insistence on individual rationality. He even affirms the unworldly nature of this rationality as a benefit when discussing the issue of intergenerational justice: the "defense of conservation does not depend on the identities of affected persons" but rather can "be made by rational persons with full knowledge of existing conditions."[15] Instead of this being a benefit, his theoretical assumptions start by erasing the real historical individuals living at the intersections of race, gender, nationality, class, geography, etc. before the rising tides even get to them. Such assumptions make it impossible to confront particularly insidious forms of "slow violence" that are often caused by and imperceptible to those at the top of global power inequalities.[16] This erasure of history is not unique to Vanderheiden but symptomatic of the liberal justice image of thought.

This effect of this ahistoricism can be seen in Vanderheiden's argument that climate change is a "unique case of global injustice" because of how it "exacerbates the global inequity that is part and parcel of the problem itself."[17] Vanderheiden goes on to argue that this and other components like the scale of climate change challenge many analytic theories of justice,[18] but never analyzes this challenge and instead proceeds as though it has no problem providing principles for a just response to climate change. But the attempt to uphold the theory blinds Vanderheiden to more critical issues. For example, if the dynamic of inequality makes climate change a unique injustice, then he should examine the literature suggesting that capitalism is a critical driver of and link between climate change and inequality. Yet capitalism goes unquestioned in the book, and market-based solutions are even suggested as a potential way of arriving at a just solution.[19] Similarly, even though Vanderheiden tries to overcome the refusal of some analytic justice theories to go beyond the nation-state and embrace cosmopolitan justice, the nation-state itself goes unexamined and is taken to be a primary actor in solving the problem. Yet the uneven political geography of the nation-state system is one of the key drivers of climate injustice.[20]

Another approach symptomatic of the liberal analytic image of thought comes from Allen Thompson and Jeremy Bendik-Keymer, who put forward what they call a "humanist view of adaptation." The question that frames their view is "what *should* become of us under global climate change?"[21] The answer is that humans should ethically flourish. But if that is the goal, then the question seems off. Perhaps it should be "what should humans *become* under climate change," to emphasize the need for a transformation to ethical flourishing. But that would also necessitate asking "how can humans become under climate change?" For it is one thing to prescribe ethics and another to say how they have been achieved. Indeed, they seem to underestimate the tension in the self at this critical moment of transformation. Though they note that "*who we are* today is not ready for this," they at the same time affirm that "who we have been got us into this mess."[22] But the problem is that we are still the ones who got us into the mess and are getting us deeper into it and living relatively well through the process, which is a serious part of *why* we are not ready for it.[23] Thus, they turn away from a vision of the human that changes in relation to the world around it, to a static rational individual around whom the world changes.

This is troubling because the framework they rely is adaptation. It is true that they point in a promising direction in turning away from the "pollutionist" and "development" views of adaptation.[24] Their own vision of adaptation "would mean changing our sense of ourselves, especially our virtues and vices, and finding what kinds of social conditions best clarify and support our dignity and decency."[25] They do not define dignity, decency, sense of self, vice, or virtue even as they point out that excellence, another of their key areas of focus, has changed dramatically in meaning over time. The meaning is thus to be found in the narrow scope of change possible given the boundary conditions set by their notion of adaptation. They justify adaptation as their framework on the basis of "the realities of anthropogenic global climate change."[26] Yet they never seem to take those realities seriously. They simply position themselves as being both "between" and "against" plans to pursue business as usual and the dire predictions of James Lovelock.[27] The problem is that Lovelock makes his dire predictions on the likelihood that humans will pursue business as usual. And yet they never show that the kind of ethics they are pursuing is suitable to a world in which many components of life, and Western life in particular, fall apart. Thus, if it is a matter of adapting to reality, we need to know what reality they think their ethics are suitable for.

A hint as to the reality they expect is given in the ethical adaptation that Thompson and Bendik-Keymer propose. In its most expansive form, it reads: "On the humanist view we offer, adaptation to global climate change is open to forms of ethical discourse that are distinct alternatives to ethical analyses concerned primarily with utilitarian calculation, international obligations, and distributional equity. In particular, the evaluative concept of excellence comes to the fore."[28] While nominally leaving the door open to other visions, they rely on the well-established theory of excellence, particularly in relation to the ancient Greek notion of *eudaimonia*. But if one is going to rely on the canonical ideas of western society, why not push them and, for example, try to extend *eudaimonia* to a concept of more-than-human flourishing? Instead, they offer two specifications of this tradition which fail to adapt it to the reality we face.

The first specification is that excellence needs to be thought in its original Greek sense, by which they mean a turn away from the individualistic moralism of contemporary virtue ethics and a return to consider human character in connection to the institutional character of human life. A promising start, but when it comes to evaluating the existing social and political institutions that have been the drivers of climate change, "the question is left open of whether protecting the status quo could be enough."[29] This means that for them it is possible that the form of human excellence that would be adaptively produced could be compatible with the general arrangement of Western society at the moment. Yet it seems hard to honestly affirm this possibility given the degree to which the carbon combustion complex is built into the structure of that society.[30] Thus, a path to a surprisingly untransformative, unadaptive ethical response is paved. Indeed, if their vision of the uniqueness of humans is that they can proactively adapt, and that the power of moral adaptation is "to calibrate our environmental character, practices, and institutions, because our *values* demand it, because we could not live with ourselves otherwise,"[31] then we find in their theory of humanist adaptation neither something very demanding nor something that we cannot live without.

Their second specification is to point out that even the original Greek sense did not include an ecological dimension. But rather than asking if it then carries a fundamental tension with being ecological, they extend it to ecology through the idea and practice of restoration. Is human superiority and the rational management of nature entailed by the practice of restoration an extension of the kinds of rationality entailed in Greek notions of excellence? This question is not asked. But they miss something more important. Just because ecology is not explicitly addressed in a given way of being or ethical system does not mean that there is not a way of relating to nature built into the thought. Indeed, all frameworks of living already entail implicit answers to how one lives in relation to the world. In an era defined by the problematic human character that has given rise to the Anthropocene, one is rightly hesitant to uncritically affirm an avowedly "humanist" view of adaptation.

In one more example of unflinching commitment to the analytic justice image of thought, John Broome's analysis of the ethics and politics of climate change relies largely on assumptions that presume little, if any, change to Western liberal capitalist conceptions individuals, government, philosophy, and markets. He assumes the primacy of the rational individual in his aspiration to help the reader in "thinking through issues of climate change for yourself."[32] His take on the science is problematically reductive in that he reads tipping points as implying only a "more than proportional"[33] relation to emissions and climate change as ultimately a "slow process" such that there is no ethical problem in assuming that political responses should take "many decades" as their timeframe for efficacy.[34] For Broome, the ethics of responding to climate change is defined by economistic reason. The primary tool he uses throughout the book is cost–benefit analysis and even when he reaches a surprisingly strict conclusion that "requires each of us to avoid emitting any greenhouse gas at all," it turns out that individuals can emit as much as they want provided they buy offsets. This assertion is made without reference to the many compelling arguments made by environmental activists, climate scientists, and others, that carbon offsets do not work.[35]

There is one moment where Broome seems to open up his analysis to a broader mode of thinking and responsiveness. Regarding questions of population and particularly how the potential for extinction fits into that question, Broome says that it raises uncertainty to such a degree that it cannot be solved by ethics. Rather, "democratic public debate is at present the only means we have of coping with this sort of uncertainty."[36] Yet even here, his assumptions undermine this possibility. Extinction is reduced to human extinction. Further, it is not really an open discussion to create alternatives but one limited to different options among the various value theories of economistic reason.[37] Further still, he presumes that the contemporary democratic system is up to the task, including that people are "well-informed participants."[38] While any national poll on climate change will show that that is not the case, it has also been shown that one need not have a lot of information out climate change to effectively participate.[39] Rather, gender, class, and race have significant impacts on how one views and responds to the issue of climate change with, for example, higher wealth being correlated with lower levels of engagement and concern.[40] If one is going to defer the question of ethically responding to catastrophic uncertainties to democracy, one then needs to ask whether a democratic system geared toward the production of wealth and providing unequal access to its mechanisms of control is up to the task. Thus, Broome's strict adherence to the analytic justice image of thought impedes the search for more effective and more ethical responses to climate change.

The challenge to ethics

Despite the dominant image of thought in analytic justice theory, it is not a homogeneous field when it comes to climate change. There are a few projects that engage the problems in this approach but fall short in overcoming them. Byron Williston's *The Anthropocene Project* exemplifies this. In the preface to his book, he uses Buñuel's *The Exterminating Angel* to suggest how far those living relatively prosperous lives are from responding to climate change. "These people are certainly frustrated by the inertia gripping them and yet—this is the key—they don't consciously pine for anything very specific outside these walls. The constricted space holding them is the outward expression of their own cramped imaginations."[41] The claim is that people cannot respond to climate change because they lack the capacity to desire and imagine a world not defined by their economic well-being. Williston then intensifies this claim by drawing on Adorno's concept of "social totality" to argue that capitalism shapes our institutions, beliefs, and desires.[42] Finally, Williston targets what he calls the "globally prosperous" because their complicity in driving climate change is the most problematic given the large impact this group of people has while remaining ideologically insulated within a liberal scientific worldview.[43] These are strong claims, though I would agree that they are largely correct. This may be the only analytic approach to climate change that takes seriously the obstacles in human consciousness in a capitalist society to building positive responses to climate change.

The problem is that having introduced these barriers, Williston quickly drops them and reintroduces a fairly standard rational enlightenment vision of humanity as the basis for a solution. Even after stating that social totality shapes our desires and beliefs, he goes on to say that we still have "the requisite convictions and values" for responding to climate change.[44] The tension between the positions goes unexplained. While Adorno uses totality in different ways, none of them suggest the ease with which Williston imagines people being persuaded to emphasize beliefs that they already have but have neglected. Indeed, with Horkheimer in the *Dialectic of Enlightenment*, Adorno analyzed the totality of the Enlightenment and reason itself as the foundation of contemporary political oppression and closed thought. Having set up the problem using this notion, he turns to argue for a solution that is "explicitly an extension of the older Enlightenment project."[45] In this way, Williston uncritically returns to the typical Enlightenment subject as the model of human thought and action.

Halfway through the introduction to his book, the notion of a constrained consciousness introduced earlier is gone, replaced by people unable to act consistently on principle because they are "seduced by consumption" and lacking "full self-control."[46] People seduced in this way just need a bit of rational persuasion, since every member of the globally prosperous class has "overriding moral reasons to treat people of the future as genuine moral subjects."[47] These people are "by nature corrigible," and can "in principle, be persuaded," to adopt the virtues that Williston argues for (justice, hope, and truthfulness) on the basis of their "reason-responsiveness."[48] While he does not claim that all humans endorse these virtues, his argument operates on the assumption that all humans are the kind of Enlightenment subjects that can be brought to them and the universal moral humanity that they envision through reason.

Williston's vision of the power of Enlightenment reason to solve climate change not only structures the interiority of the subject who responds to climate change but also the external response to the most complex parts of climate change. Generally, the complexity and certainty of climate change do not figure in Williston's understanding of climate change. His acknowledgment of it comes when he notes that the Anthropocene is a bit different than the initial Enlightenment dream.

> We can intervene like never before in natural systems, but we do not have full control over the effects of our interventions. This is because along with the discovery of our causal powers we have also learned that the systems we affect respond to our interventions in non-linear and therefore unpredictable ways, and that these responses will in turn affect human systems, also (sometimes) in non-linear ways.[49]

This scales back the Enlightenment dreams of expanding control over nature. Yet his response to the unpredictability of these natural and social feedback is that "we need to learn how to manage this state of affairs wisely."[50] It is unclear what ground Williston thinks has been conceded in the move from "control" to "management."[51] What this elides, however, is the very point of the unpredictability and non-linearity he refers to: they cannot be managed. At best, one can manage oneself or one's society to try to avoid crossing unknown tipping points and setting systemic reconfigurations into motion. Despite the mass of scientific knowledge we have concerning the climate system, in many ways it points to even greater unknowns rather than moving us closer to the requisite knowledge for managing the Earth system.

How can the tension between the initial view of climate inaction suggested by Williston and his return to a fairly standard analytic approach be reconciled? There is no reason that Buñuel and Adorno need to be in the book. After the preface, their analysis plays no real role. Thus, if it remains it may be that it persists as a hesitation, a doubt, a worry that things might not be quite as straightforward as the rest of the argument suggests. This, then, I would argue is a productive place from which to think. If climate change has called Western liberal capitalist mode of life into question, then as Adorno suggests, it likely does not leave its philosophical assumptions and culture untouched either. They too may have to go.

This is suggested by Stephen Gardiner, who explicitly states the possibility that our theories are insufficient for dealing with climate change but fails to pursue that possibility. In many ways, his *A Perfect Moral Storm* encapsulates the challenge that climate change poses. Both the article and the book share the same basic argument but with important differences. The argument is that climate change is a perfect moral storm, by which he means that three significant ethical storms—global, intergenerational, and theoretical—converge to form a particularly problematic obstacle to ethical responses and so pose a "perhaps unprecedented challenge."[52] I will begin with the most promising aspect of this argument, which is centered around the question of how these three "storms" converge.

In this convergence, the nature of the theoretical storm is critical for understanding how the three are able to come together. Gardiner's definition is worth quoting in full.

> The final storm I want to mention is constituted by our current theoretical ineptitude. We are extremely ill-equipped to deal with many problems characteristic of the long-term future. Even our best theories face basic and often severe difficulties addressing basic issues such as scientific uncertainty, intergenerational equity, contingent persons, nonhuman animals and nature. But climate change involves all of these matters and more.[53]

This explanation comes from the article, which is similar to that given in the book. What is different is that this constitutes the entire exposition of this storm in the article. That section contains one other paragraph, which only says that Gardiner will not discuss the details there. In that unelaborated format, this third storm is unnerving in its potential. It claims that we do not know how to think about the problem of climate change. In so doing, it implies that we have to start retheorizing upon radically new ground, building concepts adequate to the situation before making suggestions about what kinds of ethical and political responses might be pursued.

In not explaining or further engaging this "storm," it may be that Gardiner, at the time of writing the article, was unsure of how to handle such an unsettling argument. Taken at face value, it actually negates the other two storms. The nature and harms of both the global and intergenerational storms are elaborated on the basis of well-established theoretical models such as social-contract theory, virtue theory, utilitarianism, and cost–benefit analysis. But if the third storm is true, then the theories that form the substance of the first two storms are "inept" and thus the conclusions reached through them of little relevance. If the third storm is understood in this way, then the length of the article can be reduced to a few pages and its conclusion would be that philosophers need to get to work.[54]

On the other hand, if one insists that the analyses of the first and second storms are significant, then the meaning and impact of the third storm is significantly reduced, perhaps to the point of becoming irrelevant. This is the direction that Gardiner takes when elaborating the argument in the book. The first two storms are upheld and the third is domesticated to the point of minor relevance. The analytic justice image of thought reasserts itself through the ways in which both the third storm and the storm of climate change itself are made less intense to bring them in line with the applicability of the comfortable theories that Gardiner prefers.

In the book, the broad indictment offered by the language in the article is largely repeated but then quickly modified to what Gardiner calls "the global test." The test says that a theory or institution "must acknowledge and seek to address" the claim "that failure to address a serious global threat is a criticism of it, and a potentially fatal one."[55] If they fail to do so, they are inadequate and "must be rejected."[56] Here, a theory or institution no longer needs to address the actual global threat but simply the criticism that it would be bad not to. A philosophical standard of argumentation and evaluation becomes the object rather than climate change, the particular worldly qualities of which have evaporated into the abstraction of a "serious global threat." That he is more interested in philosophical evaluation than the actual problem becomes even clearer when he goes on to explicitly defend its high level of abstraction and to focus on whether a theory or institution is "oblivious, complacent, or even evasive" about the test's concerns as a way of deciding whether it passes the test or not. Despite having reigned in any critical potential of the test, one would still expect that having laid out such a test, Gardiner would begin with a ruthless subjection of his own philosophical method to the test. He does not do this, even though he admits that his minimal ethics approach might fail.[57] For a contrasting test that is both more concrete and more incisive, one might look to Harald Welzer's suggestion that with regard to climate change, political theory "must prove its worth in *a critique of any limitation of survival conditions for others*."[58] On whole, what started out as a theoretical storm that would invalidate many theories and institutions turns out to do nothing conclusively.

The rest of the book develops according to the analytic justice image of thought. For example, his goal is "to specify the global environmental tragedy in language that almost all morally serious people can accept, and so I will try not to beg any contentious theoretical questions."[59] Undermining his own attempt, Gardiner never clarifies what constitutes a morally serious person, instead relying on the analytic image of thought to do its work. This presumption of being "morally serious" is tied to another of the major problems that Gardiner identifies: moral corruption. This is "the idea that agents may subvert moral language and arguments for their own purposes."[60] These strategies include distraction selective attention, pandering, and delusion among others. They are generally similar to what are often classified as forms of climate denial. By individualizing (treating as self-serving) denial and calling it corruption, Gardiner carves out space for an uncorrupt moral agent inside every human. This has at least two problems. First, that if humans are not at base uncorrupt serious moral agents, then Gardiner has dramatically underestimated the problem of denial. Second, such a strategy of moral seriousness is likely to politically fail both because it is individualizing and because it makes it too easy to reduce any advocacy to naught by pointing out the hypocrisies (corruption) in the actions of any actor.[61]

The ahistorical and static nature of this moral agent is a key element of the analytic justice image of thought and finds further expression in Gardiner's analysis. Despite the kind of transformation implied even in the term climate change—it's not for nothing that one of Naomi Klein's slogans is "change or be changed"—he gestures in this direction but fails to follow through. While he points out in Chapter 1 that "climate change is likely to raise serious, and perhaps uncomfortable, questions about who we are and what we want to be," and promises to deal with this in more detail in Chapter 10, the question does not come up again in the rest of the book.[62] For comparison, Joanna Zylinska's uses a different theory of minimal ethics to look at what transformations human life should undergo through the "complex and dynamic network of relations in which 'we humans' are produced as humans."[63] For Zylinska, there is no innate human to serve as the base of analysis and so transformation is central to responding to climate change, whereas for Gardiner it difficult to think about.

Similarly, Gardiner notes the existence of tipping points and extreme scenarios that could produce "another planet."[64] Yet in each case, he finds a reason to not have to deal with them. For example, "the incremental effects of climate change are serious enough to justify substantial action without calling on catastrophic abrupt changes."[65] It is true that incremental climate change entails dire impacts, but abrupt climate change does so in an altogether more intense and uncertain way. Such tipping points present a problem for which analytical philosophy simply has no tools. One cannot conduct a cost–benefit analysis for a world in which it is possible on the one-decade timeline for coastal cities to disappear, a previously unknown pandemic to spread, or for the carbon savings through a large-scale green energy transition to get swamped by a massive methane release.

Finally, the market assumptions of the liberal justice image of thought constrain Gardiner's analysis of climate ethics. Among his "favored, and very limited, assumptions" is that "the main driver of [climate change] is the consumption behavior of agents."[66] Yet an extensive literature exists pointing to the problems of individualizing consumer choice within a capitalist system and more broadly of the destructive relation that capitalism establishes between human societies and ecosystems. This is why the term "capitalocene" has come to prominence as a way to describe the era of climate change. A case could be made for focusing on individual consumers if one targets it at the wealthiest, but Gardiner does not do this and it would go beyond his individualist assumptions to make class dynamics the basis of ethical and political analysis.[67] Thus, it is not surprising that when it comes to proposing radical cuts to carbon, Gardiner judges the extreme case of making 80% cuts by emergency Presidential decree overnight to be too unethical because of the "social and economic catastrophe for current people . . . Civil society would collapse."[68] It is enough to simply place his comment next to a one by Vinay Gupta: "Collapse means living in the same conditions as the people who grow your coffee."[69] This exposes the privilege, classism, and injustice built into Gardiner's assumptions and, generally, of the liberal justice image of thought.

Dale Jamieson's *Reason in a Dark Time* perhaps goes the furthest from within the analytic tradition in looking at how the problem of climate change exceeds that approach. Though it is tainted by nostalgia for it, it is an important beacon for analytic philosophies of climate change. Jamieson does not easily reach the conclusion that analytic theory is not equipped to think through the ethical and political implications of climate change. The book, over 20 years in the writing, is genuinely caught in a transitional moment and the author is simultaneously pulled toward thinking through the inevitability of that transition and toward the entrenched routines, thoughts, tools, and approaches that climate change unsettles. It is this very oscillation, this being caught in the middle of things, that is the book's virtue. To give a sense of this, it is worth discussing the two competing tendencies pulling at Jamieson.

The broad way that Jamieson has of speaking about the crisis that climate change has brought on is the idea that the Enlightenment is over. "It feels as though we are living through some weird perversion of the Enlightenment dream. Instead of humanity rationally governing the world and itself, we are at the mercy of monsters that we have created. This book is about how it came to this. The Enlightenment dream was a good one. Why has it failed and in the ways it has?"[70] I will put aside the questions of whether the Enlightenment really was such a solid fixture that has only now come to an end with climate change, and of what the "Enlightenment values" are that have now crumbled under the weight of climate change. What is significant is that even as he writes about its collapse, he remains deeply committed to it and unable to fully extricate himself from the building as it collapses.

This conflicted position can be seen in how he handles the issue of the novelty of climate change. On one hand, "the Anthropocene presents novel challenges for living a meaningful life"[71] that, "in moral terms would require revising our everyday understandings of moral responsibility."[72] Jamieson is aware that not only has climate change eroded the foundations of his ethical edifice but that it is part of a new situation which requires rethinking morality. Yet when it comes to doing so, he remains deeply attached to those foundations. When thinking about "revising" values in the face of novelty, his move is to lean on inherited classical and Christian virtues, rather than engaging the innovative requirement that he acknowledges is necessary, but which he never takes on.[73] Indeed, at the end of a section on how to revise morality, he says that "it would be distressing if an adequate ethics of the Anthropocene required compromising liberal ideals."[74] Those liberal ideals are the same ones rooted in the Enlightenment, which he declares over. He recognizes the problem, but finds himself unable to truly confront it, to engage the open horizon beyond the inherited concepts and ideals. This shows a man caught in his image of thought as the event erupts: he has lost his bearings and even recognizes this but is also unable to move forward and so tries to cling to them rather than let them go.

This can be seen in an even starker way in his handling of the complexity of climate change. While Jamieson is aware of some of the complexities of climate change like feedback loops and their terrifying manifestation as tipping points, he does not engage what it means for ethics and politics if there are hidden lines which, once crossed, moot human carbon reduction. Indeed, he reduces tipping points to the regular emergency questions that climate change raises.[75] Nonetheless, that does not stop the problem of complexity from putting him in a bind. On one hand, he is tempted to try and solve it according to the regular models of analytical philosophy by reducing it to a question of probability[76] or simply making climate change "the world's largest and most complex collective action problem."[77] On the other hand, he finds it necessary to admit that it goes "beyond the resources of commonsense morality"[78] and "is not a problem that conforms to our traditional models of individual morality and global justice."[79] Again, Jamieson is caught trying to take in the problem but at the same time remaining attached to approaches that cannot accommodate it.

Jamieson's analysis is thus symptomatic of climate change as an event. As climate change reshapes the world, we see the stress and tension of that reshaping in Jamieson himself. But because of how deeply entrenched he is in the analytic justice image of thought, he cannot see it as a transformation but only as an ending. With reference to Hegel, he claims "the dusk has started to fall with respect to climate change and so the owl of Minerva can spread her wings."[80] For him, the philosophical task is to understand why the attempt to prevent climate change failed. All that is left is to give a few half-hearted recommendations for abatement, mitigation, and adaptation, and hope that life will continue to be meaningful as humans "manage to live as best we can and hope that the darkest scenarios do not come to pass."[81] He knows that many analytical philosophers will reject his analysis and try and discredit him as a pessimist but only so that they can insist that the collapse he has outlined has not happened. His strength is in showing how they can only be seen as decadents, clinging to something whose time has passed. In the next section, I argue that theory has tools beyond the analytical ones which can help more productively think through the transformative event of climate change.

An ethics for the event

In the final section of this chapter, I will turn to Kierkegaard to introduce a different approach to responding to climate change. Kierkegaard writes about God rather than climate change. Yet he is instructive in that for him God erupts into life. Existing ethical responses that are universalizable and can be passed on through reason and learning are insufficient. A necessary part of responding is thus to be able to get rid of the hold such systems have on action. This is epitomized in the command that Abraham sacrifice his son, which cannot be approached in an ethical way without moving beyond systematic ethics. In this way, Kierkegaard helps us move beyond the limitations of the systematic ethics that pervade the analytic justice image of thought. This is the condition of possibility of creating vigorous responses to climate change. I will only sketch Kierkegaard's challenge to rational and universalizable ethics here as a contrast to this image of thought. The next chapter will give more detail on the specifics of Kierkegaard's approach.

In *Fear and Trembling*, Kierkegaard suggests a poetic response to the event with three aspects. That he makes philosophy poetic is key: the endeavor will be to shape and create rather than analyze, apply, and judge. First, the poet needs to go beyond everyday language to shake people up and call their attention to the uncertainty and possibility of what is happening. This helps break through the habitual thought, feeling, and social routines that obscure it. Second, the poet dramatizes the difficulty of responding to the event to help develop the aspirational energy for doing so. Finally, the poet creates and suggests multiple courses of action which are defined less by the surety of the outcome than the intensity of the transformation through which one responds.

In *Fear and Trembling*, Kierkegaard begins with a man who hears a story about god,[82] much as we hear reports about climate change. This story of god commanding Abraham to sacrifice Isaac had little meaning until his life became "fractured."[83] Fractured life awakens an uncertainty making it impossible to securely proceed as before. Today, climate change has fractured the fundamental understandings of the world to which many still cling. Indeed, the complex ways in which it confuses our previous understanding has been called "The Great Derangement."[84] When caught in uncertainty, Kierkegaard suggests that the stories of others in similar straights call to us.

The man "had but one wish, to see Abraham, but one longing, to have witnessed that event."[85] With difficulty responding to the fractures in his own life, he sought inspiration in Abraham's event. Yet he did not seek a solution and did not wish to see Abraham's reward for following through or the other glories of god. Rather, "what occupied him was . . . the shudder of the idea."[86] It makes him shudder because he is unable to understand the story of Abraham. This is the core of the event for Kierkegaard: something that ultimately cannot be comprehended because of the uncertainty and possibility that it creates. This carries him beyond the analytic image of thought, for which understanding is assumed. For Kierkegaard, one must dwell in the unknown and unsettling nature of the event and the anxiety it creates. This is the approach that Donna Haraway takes to climate change when she urges us to "stay with the trouble."

In place of understanding, Kierkegaard argues for a poetic engagement with the event to bring out the intensity of the problems it presents and the passion of the response it calls for. The poet must wrench the event away from "eternal oblivion" in which it passes "through the world as a ship through the sea, as wind through the desert, an unthinking and unproductive performance."[87] Norms, consumption habits, aspirations for wealth, the idea that nature is a resource, and other assumptions can leave us oblivious to an event like climate change or obscure it as one occurrence among others. It may be seen as a regular technical obstacle to be overcome with better engineering. The same passive effect is achieved less dramatically when we live day to day, one concern to the next, unable to give weight to an event of such scale. This is what the poet seeks to overcome in making the event come alive for a contemporary public. Haraway pushes for a similar outcome when she argues that "staying with the trouble requires learning to be truly present" rather that treating the present as a "vanishing pivot" between past and future.[88]

The poet does not give an answer but makes us feel how troubling the problem is. Kierkegaard dramatizes the problem of trying to comprehend the incomprehensible by imagining different ways in which Abraham might have carried out the sacrifice of Isaac. Abraham cannot have hesitated, he cannot have unintentionally tensed his muscles in anguish, he cannot have sacrificed himself at the last minute instead of Isaac. These poeticizations start to probe the potentiality of different ways of responding but avoid the deeply problematic and uncertain nature of the event. Each of them becomes comprehensible by showing Abraham's resistance to doing what he was commanded to do. As difficult as any of those versions of the act would be, their shortcomings call our attention to the even greater difficulty contained in the event.

The poet must also contend with active countermeasures in society that obscure the event. Kierkegaard exposes the problem of those who assume an eventless world so that they can act on the basis known results:

> When in our age we hear these words: It will be judged by the result—then we know at once with whom we have the honor of speaking. Those who talk this way are a numerous type whom I shall designate under the common name of assistant professors. With security in life, they live in their thoughts: they have a *permanent* position and a *secure* future in a well-organized state. They have hundreds, yes, even thousands of years between them and the earthquakes of existence; they are not afraid that such things can be repeated, for then what would the police and the newspapers say? Their life task is to judge . . . according to the result.[89]

Kierkegaard exposes a regime that tries to downplay the event. The combination of a secure state, a secure job, the order of a police force, and the regularized information of the media militate against the idea that something might occur which would disrupt this regime, calling into question its interpretation of the world and the security of its future. It deploys both pedestrian journalistic knowledge as well as elite scientific knowledge in conjunction with the forces of order to presumptively refute the possibility that the conditions of worldly existence could be different than assumed. It is not necessarily a regime consciously organized to resist events; it manifests itself in the routines and thoughts of those who live in such a state or have such positions. They live securely and want to judge only on the secure grounds, expectations, and calculations of that life. If an event breaks through that calls this regime into question, the standard of judging by the result is mobilized against it. This mobilization brings the inertia of an established way of life against an uncertain development, preempting its result because it is unknown. Today climate change denial is constructed through informational, economic, psychological, religious, emotional, political, and technological countermeasures, to name a few. In each case, the troubling nature of climate change is funneled into a more "manageable" mode of analysis. The analytic theory of justice is an example of this process.

In Kierkegaard's time and still in many places today, the predominant discourse of the church suppresses the event and its attendant uncertainty, instability, and anxiety. He points to the example of a preacher who preaches the story of Abraham on Sunday, and the next day condemns the man who follows Abraham's example by murdering his own son.[90] The problem is that the preacher has not thought through the event that Abraham experienced and is not able to draw out the tremendousness of it for his congregation. What is missing is the poetic valuation of the event. Kierkegaard contrasts a ready-made Christian understanding of the event with a poetic one that tries to engage it:

> We praise God's mercy, that he gave him Isaac again and that the whole thing was only an ordeal. An ordeal, this word can say much and little, and yet the whole thing is over as soon as it is spoken . . . If I were to speak about him, I would first of all describe the pain of the ordeal. To that end, I would, like a leech, suck all the anxiety and distress and torment out of a father's suffering in order to describe what Abraham suffered, although under it all he had faith.[91]

The task of the poet is to "suck all the anxiety" out of the event to enable the public to dwell in the force of it, with its uncertainty, incomprehensibility, and potentiality. The poet must first overcome the everyday ways we have of forgetting through language, routine, occupation, and so on. Then, the poet not only points out how difficult and incomprehensible the event is but also makes us interested in it, perhaps even eager to engage it.

Anxiety is a mixture of apprehension and a strong desire or concern for something. Kierkegaard calls anxiety an "ambiguous power" that both repels and draws us: "flee away from anxiety, he cannot, for he loves it; really love it, he cannot, for he flees from it."[92] We need ways to attach ourselves to the earthquakes that call us to transform ourselves, even as they disturb us. If the poet tries to instill anxiety about the event in us, it is not to drive us away from it but to enable us to confront the uncertainty of it. "[Anxiety] is altogether different from fear and similar concepts that refer to something definite, whereas anxiety is freedom's actuality as the possibility of possibility."[93] Fear has a definite object, so it is a strategic question of how to escape or overcome it. But if the object is in some way indefinite, fundamentally incomprehensible, constantly shifting, or stretches the boundaries of agency, then the simple "freedom" to act is insufficient. What is required is an open-ended freedom that engages possibility to both try to shape conditions as they emerge but also to shape the agent acting in the face of those conditions.

What energy enables this freedom and possibility to be engaged? In *Fear and Trembling*, faith is the noblest response and the reason for Abraham's greatness. Yet at the same time, the poet regularly assures us that "by no means do I have faith."[94] Thus, the poet deploys admiration, to help the public attend to noble responses to the event that can be aspired to or modeled in practice.[95] It is unlikely that anyone will do what Abraham has done. But through admiration, the poet keeps the event alive, pushes those who might not yet be engaged toward the decision to respond, and builds aspirational energy. Rather than brushing faith aside because of its difficulty, the poet shows how it is even more attractive for that.

Even the act of faith, however, is not an answer that solves a problem. Rather, it carries the problem forward because like the event, it is incomprehensible. "Even if someone were able to transpose the whole content of faith into conceptual form, it does not follow that he has comprehended faith, comprehended how he entered into it or how it entered into him."[96] Faith cannot be reduced to a concept. Just as Abraham was faced with the incomprehensibility of god's command to sacrifice Isaac, so too does Abraham become incomprehensible to others. "Humanly speaking, [the knight of faith] is mad and cannot make himself understandable to anyone. And yet, 'to be mad' is the mildest expression."[97] Faith is the affirmation of this incomprehensibility. It insists beyond the borders of reason and even breaks them down so that one does not know whether or not one is mad. But it is not just a matter of reason. Faith cannot be formulated in language or communicated, which makes speaking to others to gain one's bearings in faith impossible.[98] This highlights how far Kierkegaard's approach is from the analytic justice image of thought, which assumes that people are reasonable and have clear modes of communication, without which the system collapses.

Even if faith cannot be comprehended, it is expressed in other ways that attune us to it. Anxiety is an expression of faith that draws on the very insufficiency of reason and language to it. Put in the stark terms that Kierkegaard uses: How would one who copied Abraham know if he were truly sacrificing his son out of faith? This assurance is hard to articulate because faith resists systematization and universalization.

> [Abraham] must love Isaac with his whole soul. Since God claims Isaac, he must, if possible, love him even more, and only then can he *sacrifice* him . . . But the distress and the anxiety in the paradox is that he, humanly speaking, is thoroughly incapable of making himself understandable. Only in the

moment when his act is in absolute contradiction to his feelings, only then does he sacrifice Isaac, but the reality of his act is that by which he belongs to the universal, and there he is and remains a murderer.[99]

Abraham loves both god and Isaac with his whole soul. This traps him in a paradox because he must sacrifice one to the other. Yet anxiety is produced not only by the event of god's command but also by his inability to explain the command he is under. When Kierkegaard speaks of the universal, he is referring to the ethical state that all humans share. But Abraham becomes an exception from humanity in this regard, and his infraction against ethics (fathers should love their sons, one should not murder another) can only be seen from the view of the universal as murder. Anxiety is the strongest expression of faith in Abraham's relation to the event, because it is intensely real without being a sure, calculable relation.

The poet helps us feel the uncertainty by explaining that anxiety is anxious about its own grounds. "Be it a duty or whatever, I cannot make the final movement, the paradoxical movement of faith, although there is nothing I wish more. Whether a person has the right to say this must be his own decision; whether he can come to an amicable agreement in this respect is a matter between himself and the eternal being, who is the object of faith."[100] It comes down to a decision in which one cannot rely on others, cannot have a calculable outcome, and has no further ground to proceed upon than one's own faith. It is not a matter of willing it but of negotiating the rigors of anxiety, sounding out one's relation to the event. Kierkegaard argues that an act of faith cannot even be judged by its result.[101] This suggests that those who are comfortable with their own actions or their proposals for action on climate change might not yet have grasped the uncertainty of what is being attempted.

Anxiety is an increasingly significant symptom of climate change and has led to the production of the term "ecoanxiety." More research is being done on how events ranging from the devastation of a particular storm or drought to the unclear but impending dread of climate change as a whole affect people psychologically. Yet it is already clear that the impacts are significant.[102] Sides are already being staked out as to whether this anxiety is paralyzing, productive, or just another psychological issue to be managed. One instance stands out for its parallel to the story the poet tells in *Fear and Trembling*. *S-Town* tells the story of John B. McLemore, who wrote to *This American Life* asking them to investigate an alleged murder in his hometown Woodstock, Alabama. But what struck a number of listeners what not the story of the murder but McLemore's obsession with climate change. Some listeners became obsessed with McLemore and the passion that drove him.[103] His story sparked a discussion of psychology and anxiety because of the strength of that obsession and the way it drew people in through the story. Despite his clear analysis of climate change and logical justification of his position, he nonetheless appeared to many as "crazy." He assembled lectures, data, and slideshows and yet was unable to persuade even those who agreed that climate change was real of how truly significant it was, even losing a close friend over the issue. He also committed suicide in the middle of the production of the series, citing ecological grounds in his note.

Climate change is most often comprehended through data. While this data was important for McLemore, he could not make others see the same message in it that he saw. When people react to his anxiety and begin to feel anxiety themselves, it is a result of his passion and intensity rather than any fact that he brings forward. Brian Reed, the journalist who tells McLemore's story, is the poet in this case. He too is a bit obsessed with the strange and intriguing character of McLemore and though he does address the issue of climate change from time to time, he always pushes it aside despite McLemore's insistence. This may, however, unintentionally serve the poetic cause. Listeners encounter Reed as someone unwilling to engage the troubling nature of climate change because it is too difficult, or distracting, or impossible. In contrast, McLemore stands out all the more as someone deeply immersed in its disastrous possibility, as someone through whom we can recognize the event and be stimulated by it. Some may criticize this positive valuation and reduce it to the argument that it amounts to a call for people to kill themselves. That solution is too easy and misses the anxiety. Nonetheless, the possibility of suicide should not become an object of fear used to dismiss the case of McLemore, just as Kierkegaard's poet does not "fear arousing a desire in people to be tried as Abraham was."[104] To limit possibility before one has even begun to deal with the situation is already a refusal to deal with the troubling anxiety of the event.

Finally, the poet demonstrates the concrete effects of spiritual engagements with the event. Despite its seeming divine orientation, the object of faith is this world.[105] "If [Abraham's] faith had been only for a life to come, he certainly would have more readily discarded everything in order to rush out of a world to which he did not belong . . . But Abraham had faith specifically for this life—faith that he would grow old in this country, be honored among the people, blessed by posterity, and unforgettable in Isaac, the most precious thing in his life."[106] Faith is a commitment to and an immersion in the world. This might seem unnecessary, but as Bruno Latour has pointed out, many political materialists and even scientists are themselves not really that materialist and are even otherworldly in their understanding of this world.[107] Climate change can interrupt this abstracted living to catalyze a new connection to Earthly life.

Worldliness itself is not something abstract but manifests in all kinds of actions. The poet praises the tone of voice and the timing of the knight of faith. He describes the knight of faith: "He is solid all the way through. His stance? It is vigorous, belongs entirely to finitude . . . every time one sees him participating in something particular, he does it with an assiduousness that marks the worldly man who is attached to such things."[108] Kierkegaard further depicts the meal this knight hopes for and the poorer meal he enjoys just as much, the way he smokes his pipe, observes the street, and talks with a stranger. How we carry out even small daily acts matters. Today, the manner in which many people behave suggests a mind elsewhere than with the world. Kierkegaard compares the knight of faith to the bourgeois philistine, an entirely materialistic person who has never given a thought to the spiritual and cares only for the amassment and pleasure of worldly possessions. A difference cannot be spotted between them until one notices that the knight enjoys minimalism and simple pleasures just as deeply as rich and exclusive ones. It is not the material world that the philistine enjoys, but the idea of abundance, which divides him from it. Alternatively, through the faith that "all things are possible," the knight of faith attains a "security that makes him delight in it as if finitude were the surest thing of all."[109] Faith is a movement through which one becomes convinced that what they desire most is impossible but then continues to live on the basis of possibility itself as it unfolds in the world.

Events are upsetting, interruptive, and fracture life. This is the case with climate change, which floods homelands and desiccates agricultural land. Yet numb to such effects, many living off the destructive bounty of carbon capitalism refuse to take seriously and accommodate themselves to the idea of a "decreased standard of living." Arguments against doing so are made by a number of analytical justice theorists. They refuse a material existence in their attachment to an ideal. The knight of faith is just as happy in a world with the kind of constrained consumption opportunities and reduced mobility that would go along with cutting carbon emissions, as one without such limitations. The knight might have idealized visions of environmental progress but does not fall apart when they fail to materialize. Kierkegaard calls this movement "absurd" because the knight knows how little sense belief makes, always feels the anxiety of its impossibility, and yet believes in possibility anyway.

Kierkegaard is aware that not everyone can make the movement of faith. Indeed, the task of the poet is in part to elevate this as an aspirational ideal. Thus, he contrasts it with what he calls "spiritual trial," which has a number of forms such as irony, infinite resignation, and humor that Kierkegaard thinks are productive. Edna and Howard Hong describe this movement as "the struggle and the anguish involved in venturing out beyond one's assumed capacities or generally approved expectations."[110] Spiritual trial works to engage the troubling nature of the event, undertaking uncomfortable experiments, but without fully coming to terms with it. This is not bad and would certainly be a step beyond where most people are today. Even people who have become depressed about climate change and who no longer believe that a good outcome is possible may not have begun working through its troubling nature. Kierkegaard suggests that such people are just at the point where they are ready to begin spiritual trial.[111] Even the lesser engagement of spiritual trial can be enticing insofar as it enables us to dwell in the problems of the event and build a response to them that gives some repose, because spiritual trial can lead to "peace and rest and comfort in the pain."[112] Someone undergoing spiritual trial might not enjoy living a life under climate change, but they could still console themselves with an ideal of what humanity and the world could have been given the contingent possibilities of history.

Still, the poet dangles faith as a lure beyond this repose. For spiritual trial is "incommensurabl[e] with actuality."[113] It does not manage to remain grounded and worldly, despite is experimental forays into responsiveness. Those who enact the movement of faith, on the other hand, "exist in such a way that [their] contrast to existence constantly expresses itself as the most beautiful and secure harmony with it."[114] Committing to the potentiality and problems of the event brings about a state of being that affirms the unruly world of finitude. This is admirable and astounding. Though dozens if not hundreds of approaches, understandings, interpretations, and solutions to climate change have been put forward, none suggest living in harmony with it. Significantly, Kierkegaard does not point to a world that is harmonious but to a subject who manages to bring themselves to live in harmony with a world that is uncertain, tragic, and nonetheless beautiful. So, the poet tries to tempt us to go further, even though he continues to insist that he cannot, as much as he would like to. But ultimately, "whether the single individual actually is undergoing a spiritual trial or is a knight of faith, only the single individual himself can decide."[115] Whether one breaks out of spiritual struggle and into faith has no ground outside oneself, which is why it resists universalization. This groundlessness produces the intensity of the struggle.

Some might say that the poet is making too big of a deal out of things: even if climate change is a serious global problem, it is not beyond comprehension and even has a number of solutions. It is simply a matter of subsidies for research, development of green technologies, carbon taxes, perhaps a restructuring of politics to get rid of corporate money, a binding global treaty, or, some would argue, a more serious engagement with geoengineering. Many of these are the suggestions given by the analytic justice thinkers discussed in the first two sections of this chapter. Such responses gloss over the trouble. They make prescriptions based on an idealized world of anthropocentrism, human rationality, and liberal individuality rather than the existing one, which does not conform to that vision of reason and is not necessarily hospitable to humans. Climate change is an event that shatters such idealizations. The poet calls our attention to the ways in which it does so, trying to pull our thoughts and actions away from such mental routines and toward an uncertain but passionate responsiveness and worldliness. The point is to transform ourselves at least as much as the world is being transformed. But even as this engagement calls attention to the issue, it does not yet deal with the subjective problems of "staying with the trouble." The next chapter puts Kierkegaard into closer conversation with Haraway on how to do so.

Notes

1 For a critique of the capacity of analytic justice theory to deal with environmental issues, see Read, "Why the Ecological Crisis Spells the End of Liberalism."
2 For similar arguments that deal with parts of what I'm calling the analytic liberal justice image of thought, see Bonneuil and Fressoz, *The Shock of the Anthropocene*, chapter three and Wainwright and Mann, *Climate Leviathan*, chapter four.
3 Trojanow, *The Lamentations of Zeno*, 108.
4 Shue, *Climate Justice*, 265.
5 Ibid.
6 Ibid., 9.
7 Ibid., 9–10.
8 Welzer, *Climate Wars*, 171–2.
9 Shue, *Climate Justice*, 25–6.
10 Klein, "Capitalism vs. the Climate."
11 For a critique of sociocentric approaches that rely on a gradualist vision of the science, see chapter one of Connolly's *Facing the Planetary*.

12 Vanderheiden, *Atmospheric Justice*, 44.
13 Ibid.
14 Ibid.
15 Ibid., 142.
16 Nixon, *Slow Violence*.
17 Vanderheiden, *Atmospheric Justice*, xiv. See p. 78 as well.
18 Ibid., xiv–xviii, 79. These are a couple examples, though this idea comes up throughout the book.
19 Ibid., 140.
20 Welzer, *Climate Wars*.
21 Thompson and Bendik-Keymer, *Ethical Adaptation*, 1.
22 Ibid., 15.
23 Though it is not clear, I read their "we" in terms of the relatively wealthy classes, mostly located in Europe and the Anglo societies.
24 Thompson and Bendik-Keymer, *Ethical Adaptation*, 5–7.
25 Ibid., 2.
26 Ibid.
27 Ibid., 1.
28 Ibid., 8.
29 Ibid., 9.
30 Oreskes and Conway, *The Collapse of Western Civilization*.
31 Thompson and Bendik-Keymer, *Ethical Adaptation*, 9.
32 Broome, *Climate Matters*, 9.
33 Ibid., 34–5.
34 Ibid., 190.
35 See Monbiot, "Paying for Our Sins"; Anderson, "The Inconvenient Truth of Carbon Offsets."
36 Broome, *Climate Matters*, 191.
37 Ibid., 183–6.
38 Ibid., 186.
39 Read, "What Do People Know About Global Climate Change?"
40 Norgaard, "Climate Denial and the Construction of Innocence."
41 Williston, *The Anthropocene Project*, vii.
42 Ibid., viii.
43 Ibid.
44 Ibid.
45 Ibid., 22.
46 Ibid., 9.
47 Ibid., 8.
48 Ibid., 9–10.
49 Ibid., 11.
50 Ibid.
51 The literature criticizing the domination, control, and management of nature is large. In the context of climate change, see for example Luke, "Reconstructing Social Theory and the Anthropocene."
52 Gardiner, *A Perfect Moral Storm*, 7.
53 Gardiner, "A Perfect Moral Storm," 407.
54 Though he lacks the stark potential of Gardiner's third storm, John Nolt goes further in arguing that anthropocentrism and the temporality of climate change severely limit the usefulness of analytic justice theory in his article "Long-Term Climate Justice."
55 Gardiner, *A Perfect Moral Storm*, 218.
56 Ibid.
57 Ibid., 220.
58 Welzer, *Climate Wars*, 163.

59 Gardiner, *A Perfect Moral Storm*, 5.
60 Ibid., 46.
61 Ghosh, *The Great Derangement*, 132–5.
62 Gardiner, *A Perfect Moral Storm*, 31.
63 Zylinska, *Minimal Ethics for the Anthropocene*, 13.
64 Gardiner, *A Perfect Moral Storm*, 101, 113, 264, 359.
65 Ibid., 101.
66 Ibid., 58.
67 Chancel and Piketty, "Carbon and Inequality."
68 Gardiner, *A Perfect Moral Storm*, 20.
69 Gupta, "Time to Stop Pretending."
70 Jamieson, *Reason in a Dark Time*, 1.
71 Ibid., 8.
72 Ibid., 168–9.
73 Ibid., 186.
74 Ibid., 176.
75 Ibid., 221.
76 Ibid., 102.
77 Ibid., 162.
78 Ibid., 6.
79 Ibid., 8.
80 Ibid., 1.
81 Ibid., 10.
82 Kierkegaard refers to both God and god in his writings, sometimes arguing that it is just a name without importance. In this book I follow the latter orthography.
83 Kierkegaard, *Fear and Trembling*, 9.
84 Ghosh, *The Great Derangement*.
85 Kierkegaard, *Fear and Trembling*, 9.
86 Ibid.
87 Ibid., 18.
88 Haraway, *Staying with the Trouble*, 1.
89 Kierkegaard, *Fear and Trembling*, 62–3.
90 Ibid., 28–9.
91 Ibid., 52–3.
92 Kierkegaard, *The Concept of Anxiety*, 44.
93 Ibid., 42.
94 Kierkegaard, *Fear and Trembling*, 32.
95 Ibid., 38.
96 Ibid., 7.
97 Ibid., 76.
98 Ibid., 114–15.
99 Ibid., 74.
100 Ibid., 51.
101 Ibid., 62–7.
102 Clayton, et al., *Mental Health and Our Changing Climate*.
103 Green, "The Existential Dread of Climate Change." See also Salamon, "What S-Town Gets Wrong About Climate and Mental Health"; Matthews, "Why John B.'s Climate Obsession in *S-Town* Is So Unsettling."
104 Kierkegaard, *Fear and Trembling*, 53.
105 This connection is suggested by Gilles Deleuze in *Cinema*. For a reading of how Kierkegaard uses the comic to build worldly attachment, see Tønder, "Comic Rules."
106 Kierkegaard, *Fear and Trembling*, 20.
107 Latour, *Down to Earth*, 58–70.

108 Kierkegaard, *Fear and Trembling*, 39.
109 Ibid., 46, 40.
110 Ibid., 343n14.
111 Ibid., 48.
112 Ibid., 45.
113 Ibid., 51.
114 Ibid., 50.
115 Ibid., 79.

Bibliography

Anderson, Kevin. "The Inconvenient Truth of Carbon Offsets." *Nature* 484, no. 7392 (April 2012). doi:10.1038/484007a.

Bonneuil, Christophe, and Jean-Baptiste Fressoz. *The Shock of the Anthropocene: The Earth, History and Us*. Translated by David Fernbach. New York: Verso, 2017.

Broome, John. *Climate Matters: Ethics in a Warming World*. New York: Norton, 2014.

Chancel, Lucas, and Thomas Piketty. "Carbon and Inequality: From Kyoto to Paris." *Paris School of Economics*, November 3, 2015. http://piketty.pse.ens.fr/files/ChancelPiketty2015.pdf.

Clayton, Susan, Christie Manning, Kirra Krygsman, and Meighen Speiser. *Mental Health and Our Changing Climate: Impacts, Implications, and Guidance*. Washington, DC: American Psychological Association and ecoAmerica, 2017. www.apa.org/news/press/releases/2017/03/mental-health-climate.pdf.

Connolly, William E. *Facing the Planetary: Entangled Humanism and the Politics of Swarming*. Durham: Duke University Press, 2017.

Deleuze, Gilles. *Cinema 2: The Time Image*. Translated by Hugh Tomlinson and Robert Galeta. Minneapolis: University of Minnesota Press, 1989.

Deleuze, Gilles. *Difference and Repetition*. Translated by Paul Patton. New York: Columbia University Press, 1994.

Gardiner, Stephen M. "A Perfect Moral Storm: Climate Change, Intergenerational Ethics and the Problem of Moral Corruption." *Environmental Values* 15, no. 3 (August 2006): 397–413. doi:10.3197/096327106778226310.

Gardiner, Stephen M. *A Perfect Moral Storm: The Ethical Tragedy of Climate Change*. New York: Oxford University Press, 2011.

Ghosh, Amitav. *The Great Derangement: Climate Change and the Unthinkable*. Chicago: University of Chicago Press, 2016.

Green, Emily. "The Existential Dread of Climate Change." *Psychology Today*, October 13, 2017. www.psychologytoday.com/us/blog/there-is-always-another-part/201710/the-existential-dread-climate-change.

Gupta, Vinay. "Time to Stop Pretending." Talk Given at Uncivilization Dark Mountain Festival, Llangollen, Wales, May 29, 2010. www.youtube.com/watch?v=EkQCy-UrLYw.

Haraway, Donna. *Staying with the Trouble: Making Kin in the Chthulucene*. Durham: Duke University Press, 2016.

Jamieson, Dale. *Reason in a Dark Time: Why the Struggle Against Climate Change Failed—Wand What It Means for Our Future*. New York: Oxford University Press, 2014.

Kierkegaard, Søren. *The Concept of Anxiety*. Edited and Translated by Reidar Thomte in collaboration with Albert B. Anderson. Princeton, NJ: Princeton University Press, 1980.

Kierkegaard, Søren. *Fear and Trembling*. Edited and Translated by Howard V. Hong and Edna H. Hong. Princeton, NJ: Princeton University Press, 1983.

Klein, Naomi. "Capitalism vs. the Climate." *The Nation*, November 9, 2011. www.thena tion.com/article/capitalism-vs-climate/.

Latour, Bruno. *Down to Earth: Politics in the New Climatic Regime*. Translated by Catherine Porter. Medford, MA: Polity Press, 2018.

Luke, Timothy W. "Reconstructing Social Theory and the Anthropocene." *European Journal of Social Theory* 20, no. 1 (2016): 80–94. https://doi.org/10.1177/1368431016647971.

Matthews, Susan. "Why John B.'s Climate Obsession in *S-Town* Is So Unsettling." *Slate*, April 13, 2017. www.slate.com/blogs/browbeat/2017/04/13/why_john_b_mclemore_s_obsession_with_climate_change_in_s_town_is_so_unsettling.html?via=gdpr-consent.

Monbiot, George. "Paying for our Sins." *The Guardian*, October 18, 2006. www.theguar dian.com/environment/2006/oct/18/green.guardiansocietysupplement.

Nixon, Rob. *Slow Violence and the Environmentalism of the Poor*. Cambridge: Harvard University Press, 2011.

Nolt, John. "Long-Term Climate Justice." In *Climate Justice: Integrating Economics and Philosophy*, edited by Ravi Kanbur and Henry Shue, 230–46. New York: Oxford University Press, 2019.

Norgaard, Kari Marie. "Climate Denial and the Construction of Innocence: Reproducing Transnational Environmental Privilege in the Face of Climate Change." *Race, Gender & Class* 19, no. 1–2 (2012): 80–103.

Oreskes, Naomi, and Erik M. Conway. *The Collapse of Western Civilization*. New York: Columbia University Press, 2014.

Read, Daniel, Ann Bostrom, M. Granger Morgan, Baruch Fischoff, and Tom Smuts. "What Do People Know About Global Climate Change? II. Survey Studies of Educated Lay People." *Risk Analysis* 14, no. 6 (1994): 971–82.

Read, Rupert. "Why the Ecological Crisis Spells the End of Liberalism." *Green House Think Tank*, Summer 2011. www.greenhousethinktank.org/files/greenhouse/home/post_growth_commonsense_inside.pdf.

Salamon, Margaret Klein. "What S-Town Gets Wrong About Climate and Mental Health." *The Climate Psychologist*, May 5, 2017. http://theclimatepsychologist.com/what-s-town-gets-wrong-about-climate-and-mental-health/.

Shue, Henry. *Climate Justice: Vulnerability and Protection*. Oxford: Oxford University Press, 2014.

Thompson Allen, and Jeremy Bendik-Keymer. *Ethical Adaptation to Climate Change: Human Virtues for the Future*. Cambridge: The MIT Press, 2012.

Tønder, Lars. "Comic Rules." *Theory and Event* 18, no. 2 (2015). https://muse.jhu.edu/article/578635.

Trojanow, Ilija. *The Lamentations of Zeno*. Translated by Philip Boehm. New York: Verso, 2016.

Vanderheiden, Steve. *Atmospheric Justice: A Political Theory of Climate Change*. New York: Oxford University Press, 2008.

Wainwright, Joel, and Geoff Mann. *Climate Leviathan: A Political Theory of our Planetary Future*. New York: Verso, 2020.

Welzer, Harald. *Climate Wars: Why People Will Be Killed in the Twenty-First Century*. Translated by Patrick Camiller. Malden, MA: Polity Press, 2012.

Williston, Byron. *The Anthropocene Project: Virtue in the Age of Climate Change*. New York: Oxford University Press, 2015.

Zylinska, Joanna. *Minimal Ethics for the Anthropocene*. Ann Arbor: Open Humanities Press, 2014.

2 Impassioned trouble
Committing to the event

The previous chapter discussed the dominant philosophical and ethical approaches to climate change and kinds of political solutions they promote. These approaches were critiqued as being overly restrictive in their focus on rationalized liberal ethics applied to climate change. Kierkegaard's approach to ethics was presented as an alternative that accommodates the unsettling novelty of events like climate change while enabling new kinds of responsiveness. Similarly, in *Don't Even Think about it*, George Marshall argues that while reason, including scientific knowledge, is essential for alerting us to climate change, it is insufficient for pushing us to take action. Something else is needed to "galvanize our emotional brain into action."[1] With the tendencies and limitations of human psychology in mind, he turns to religion. This might seem a counterintuitive move, given how the religion/science divide has problematically influenced public discussions of climate change. Yet Marshall is not suggesting that climate change be treated like religion but rather that we should draw on the features of great religions to see how they might be "mobilized to create sacred values around climate change."[2] He rejects the word belief because of the religious frame it carries in favor of "conviction." His argument is that approaching climate change in this way will both attract value-oriented skeptics and motivate those who agree that it is an important issue but currently take insufficient action. William E. Connolly goes beyond the naming of particular values when he reads climate change through the myth of Job to engage the spiritual dimension of life in relation to climate change.[3] In turn, I ask what it would mean to believe in climate change.

Several arguments are made against believing in climate change. For example, that it undermines the role of science by skipping over the reliability of empirical data. Or that belief is too arbitrary to be able to forge political connections as part of a movement. Or even that it is a flight into the otherworldly. But even Marx in his trenchant criticism of religion and otherworldliness insists that such beliefs are material, a real opium for a suffering world. And as the last chapter demonstrated, for Kierkegaard, belief is a commitment to material existence in this world. Indeed, my argument is that the problem is on the side of a routinized and torpid materialism. Despite decades of alarming stories and increasingly dire predictions, business as usual continues. Belief, for Kierkegaard, is a matter of passion and what is needed today is an affective infusion into political responses to climate change to match the urgency of the situation.

Another problem that might be raised is that belief is too individualistic. Indeed, Kierkegaard's work is often read as reinforcing a radical individualism. If this were the case, it would put him on the side of the liberal individualist thinkers critiqued in the last chapter. But two things prevent such a reduction. First, Kierkegaard was well aware of the fractures within the self and the dissimulation involved in presenting oneself to society as a cohesive individual. In one journal entry he writes: "There is—and this is both the good and the bad in me—something spectral about me, something that makes it impossible for people to put up with me every day and have a real relationship with me. Yes, in the light-weight cloak in which I usually appear, it is another matter. But at home it will be evident that basically I live in a spirit world."[4] This matter is complicated by the fact that Kierkegaard used both his own name and pseudonyms to give expression to what might be called his multiple self.[5] In addition, this internal multiplicity cannot be compressed into the form of an individual; there is the fact that the thrust of much of Kierkegaard's thought is the insufficiency of the self. As I argue in this chapter, his focus on how we realize and respond to our dependence on forces beyond ourselves is critical to thinking about how to respond to climate change.

Kierkegaard thus helps us see climate change as a spiritual event breaking into the world and the extent to which a spiritual response is called for. Even though events interrupt life and carry a number of dangers, Kierkegaard views them as essential for developing one's spiritual disposition and existential stance in the world. To do so, I will put Kierkegaard's thought in conversation with Haraway's notion of "staying with the trouble."[6] In doing so, I will develop two sides of the notion of "impassioned trouble." First, it is easy enough to say and know that something is troubling but another to make that trouble felt. One key element of Kierkegaard's approach is that it draws out the intensity and suffering of the event, and in this way helps us impassion the trouble. Second, once the trouble is felt, Kierkegaard seeks to respond to it with belief, which for him is an extreme form of passion. In this way, he enables us to commit to the troubling event of climate change and the world that gives rise to it.

This chapter will proceed in two sections. The first picks up where the end of the previous chapter left off. There, Kierkegaard used the poet to break through the habitual thought, feeling, and social routine that obscure the event and to dramatize aspirational responses to it. But faith need not be achieved through a leap. Kierkegaard also theorizes the event as a hypostatized moment of possibility that awakens despair. Kierkegaard develops a number of techniques to work through despair and reorganize the self to better respond to the event. This suggests that Donna Haraway's notion of staying with the trouble may need to be broadened to include a more serious engagement with climate despair. In the second, Kierkegaard takes up the event as a coming-into-existence which has the potential to existentially change those who experience it. This change brings about a shift from an understanding-centric experience of the world to one infused with a pathos for the process of coming into existence. This section closes with a discussion of how Kierkegaard tried to become an event at the end of his life and whether Pope Francis issues a similar challenge in "Laudato Si'."

The nonevent and despair: how to stay with the trouble of a conditioned existence

Haraway sees climate change or, following Stengers, the intrusion of Gaia, as an event. It is momentous, dangerous, and tied to other frightening events like the Sixth Great Extinction. This is why her approach to climate change is focused on "staying with the trouble." Trouble has multiple meanings for her. On one hand, it points to the tasking of making trouble in the sense of stirring up responsiveness while also settling troubled waters.[7] On the other hand, it is an immersive presence in the time that is currently unfolding. "Staying with the trouble requires learning to be truly present, not as a vanishing pivot between awful or Edenic pasts and apocalyptic or salvific futures, but as moral critters entwined in myriad unfinished configurations of places, times, matters, meanings."[8] Staying with the trouble is a way of focusing on the uncertainty and novelty of this eventual moment. It has no necessary trajectory but is fundamentally open. But Haraway also does an excellent job of emphasizing the materiality of mortal finitude while also drawing out the richness of interconnected beings and ways of being. The event of climate change is also not simple and singular then but expressed unevenly across these variegated sets of connections.

These basic meanings of trouble are expanded and clarified throughout the text. The task is sharpened as the "core" of the concept is put forward: "Relays, string figures, passing patterns back and forth, giving and receiving, patterning, holding the unasked-for pattern in one's hands, response-ability."[9] This inheriting connections and building new ones as a game of "multispecies becoming-with" goes to the heart of what it means to inherit material conditions that one may fear and resent while at the same time drawing on the positive linkages within them and looking for new ones that can be forged to hand off a more just and livable pattern. There is no doubt that Kierkegaard could benefit from a strong infusion of this materialist symbiotic approach. Still, one point of connection that they share is the attempt to "make present . . . the world in its sheer non-one-selfness."[10] While Haraway does so on a materialist level, her approach can benefit from how Kierkegaard does so on a spiritual one. Haraway's analysis is full of intricate and inspiring examples of the multispecies collaborations that give vitality, joy, and trouble to Earthly existence. For all of these specific examples, however, the general phrase and idea that comes up over and over to give multispecies becoming-with its troubling bent is "living and dying" together. As an empirical description of life on Earth, it is perfectly accurate. But as a phrase for this event, it falls flat and flattens out the trouble we are in. Indeed, after reading it the fiftieth time, living and dying blend to an everyday ease. This is not a stylistic problem of repetition but a symptom of a deeper theoretical commitment.

It is not just life and death but the intensities driving attachment to them that Haraway undercuts. She is right to insist on focusing on the present, but her refusal of past and future is tied to a more significant refusal of hope and despair. "[W]e succumb to despair or hope and neither is a sensible attitude. Neither despair nor hope is tuned to the senses, to mindful matter, to material semiotics, to mortal earthlings in thick copresence. Neither hope nor despair knows how to teach us to 'play string figures with companion species.'"[11] It may well be true that these affects do not teach us how to play string figures, but they might help us understand why doing so matters or even give us the energy to do so. Haraway is also right to point out that there are terrible and terrifying forms of hope and despair, many of which are mobilized in relation to climate change. But China Miéville suggests a different view on these while insisting on the present. "Earth: to be determined. Utopia? Apocalypse? Is it worse to hope or to despair? To that question there can only be one answer: yes. It is worse to hope or to despair. Bad hope and bad despair are mutually constitutive."[12] Bad versions of utopia and apocalypse, of hope and despair, which rely on each other as opposition, need to be rejected. They form a trap that seems to present options but really works to prevent an honest assessment of the present and the generation of energies to transform it. It is significant that Miéville draws on some of the same science fiction authors like Le Guin that Haraway does. But instead of just reading them to think about connections in the present, he reads them for the future as well. There is a good way to hope and despair. "Pessimism has a bad rap among activists, terrified of surrender. But activism without the pessimism that rigor should provoke is just sentimentality. There is hope. But for it to be real, and barbed, and tempered into a weapon, we cannot just default to it. We have to test it, subject it to the strain of appropriate near-despair."[13] The problem with Haraway is not that she falls into a sentimental version of hope but that she strips away both hope and despair and the way they can work productively together. The world is in trouble and staying with it needs to include feeling it, even if it does provoke despair.

Kierkegaard can help us engage despair while avoiding the "destructive" and "sublime" versions of it that worry Haraway.[14] He engaged the affect of despair deeply yet without succumbing to "abstract futurism" or concluding that "the apocalypse really is nigh."[15] For Kierkegaard, engaging despair in the right way does not lead to resignation but to possibility. As psychologist Margaret Klein Salamon argues in connection to climate change, "despair should be viewed as a halfway point to action; it can be part of a process of coming to terms with climate truth and dedicating oneself to productive action."[16] Indeed, failing to engage the felt emotion of despair is a way of denying climate change.[17] Despair is thus a critical component of staying with the trouble of climate change such that it is truly felt in a way that energizes responses. This is what I call impassioned trouble.

Kierkegaard argues for the productivity of despair in *The Sickness unto Death*. The sickness unto death is despair. "Literally speaking, there is not the slightest possibility that anyone will die from this sickness or that it will end in physical death. On the contrary, the torment of despair is precisely this inability to die . . . When the danger is so great that death becomes the hope, then despair is the hopelessness of not even being able to die."[18] Kierkegaard links despair to the desire for an event (death) that will not occur. Despair is being in a troubled position and wanting a way out but knowing that no simple way out is possible. This is a good way of understanding those who Haraway criticizes for being in despair, those who are "actually working and playing for multispecies flourishing with tenacious energy and skill, while expressing an explicitly 'game over' attitude that can and does discourage others."[19] Haraway views the despair as a problematic and excess disposition to be excised from the productive activity. But what if it is this despair that is not in excess but a key element shaping and motivating that activity? Would they still feel despair in this way if they simply accepted the situation as game over and did nothing but enjoy their lives before the effects of climate change really take hold?[20]

What would it mean for there to be an event that we longed for, oriented ourselves toward, lived for and yet that would not occur? We have known about climate change for decades, have spent much longer contributing to it, and have begun to try to limit it, yet it does not occur. Are we at the beginning of it? Where is its midpoint? Once we feel the harshest effects, will that at least give us the relief that now the event is over and we can set ourselves to dealing with the new state of the world? Unfortunately, it seems that none of these are the case. The uncertainty and temporal dispersion of the event, the effects of which will continue long after the centuries-long life of the carbon injected into the atmosphere today is gone, deprive us of the ability to contain it as an occurrence. It should not be surprising that people who have spent their lives researching or campaigning should at some point feel themselves overcome by an event that emerges in a way beyond anything they will ever be able to experience. And yet they feel it, it weighs on them, they become sick of it. How does suffering through the simultaneous desire for an end to climate change, even if it is the apocalypse, and the knowledge that such an end will not occur help build actual responsiveness to the event?

Kierkegaard orients himself toward the necessary yet unknowable event of death. In its simple everyday sense, death is a "minor event," but death as the condition that haunts life becomes a sickness.[21] Becoming conscious of our mortality can lead to a desire for death to be a simple end. And yet it does not take that restricted form of a momentary occurrence but rather coexists with life and so becomes an event which will not simply occur. If death is the desired event, despair is the intensification of suffering which results from the event not occurring. This non-event ensnares life, transforming the existential character of death. "It is in this last sense that despair is the sickness unto death, this tormenting contradiction, this sickness of the self, perpetually to be dying, to die and yet not die, to die death. For to die signifies that it is all over, but to die death means to experience dying."[22] To be in despair is to suffer death without actually dying, to be caught in a catastrophe that will not end and that death cannot solve. Considered as an event that will not occur, death becomes despair, a constant weight on life.

The connection between how we relate to death and how we relate to climate change has been drawn by many, though Jonathan Franzen gives a particularly helpful version.

> Given a choice between an alarming abstraction (death) and the reassuring evidence of my senses (breakfast!), my mind prefers to focus on the latter. The planet, too, is still marvelously intact, still basically normal—seasons

changing, another election year coming, new comedies on Netflix—and its impending collapse is even harder to wrap my mind around than death. Other kinds of apocalypse, whether religious or thermonuclear or asteroidal, at least have the binary neatness of dying: one moment the world is there, the next moment it's gone forever. Climate apocalypse, by contrast, is messy. It will take the form of increasingly severe crises compounding chaotically.[23]

He is trying to explain how denial, or at least the lack of climate responsiveness, is tied to how one engages suffering in the world, how one stays with the trouble. Leaving death as a fact keeps it abstract in the same way as simply treating climate change like a generic apocalypse keeps it abstract. Here we see the difference between the bad pessimism that Haraway rightly rejects and the good pessimism that Miéville tries to get us to attend to. One can be sensuously immersed in the material world (elections, breakfast, laughter, seasons) and still too abstract in keeping out the pessimistic elements that gravitate toward despair. For Kierkegaard, those who treat death abstractly are in greater despair than those who suffer from it in the realization of their mortal condition. They are not yet conscious of the despair they are in and so have not yet begun to work through it. As Clive Hamilton points out, brief and abstract reminders of death tend to result in retreat into materialistic and individualistic behavior, while for more sustained and conscious engagements, "the evidence suggests . . . a salutary effect and contribute to a shift in value orientation that is both more mature and more protective of the environment."[24]

Once despair is engaged, it becomes clear that it is not entirely negative. It sustains as much as it tortures, for "the dying of despair continually converts itself into a living."[25] The desire to escape despair is transformed by the realization that it has no escape, no easy solution. Accepting the most horrible apocalypse does not end a catastrophic situation. The sickness cannot be solved by death and does not itself result in death. "The inability of despair to consume him is so remote from being any kind of comfort to the person in despair that it is the very opposite. This comfort is precisely the torment, is precisely what keeps the gnawing alive and keeps life in the gnawing, for it is precisely over this that he despairs . . . that he cannot consume himself, cannot get rid of himself, cannot reduce himself to nothing."[26] The positive attachment switches from comfort that despair will not result in death to comfort in the vitality of suffering. This shift establishes and maintains a relation to the event, giving body and strength to it. Climate change won't actually end the world and the crises that it sets in motion are a source of ongoing vitality as the world keeps living. Kierkegaard calls this the "provocativeness" of despair.[27] Such provocation suggests that those who work to respond to climate change while despairing over it may receive their energy for this responsiveness from despair.

Despair, even when it is about an event, is rooted in the self. Fantasies of the world on fire or submerged in water, or of green techno-utopias are not about climate change but about the subjects who dream them up. "To despair over oneself, in despair to will to be rid of oneself—this is the formula for all despair."[28]

Despair is an improper relation to the self, an attempt to forgot one's actual condition. To understand despair, we need to understand the self. For Kierkegaard, the self is composed of both internal and external relations. Internally, the human is a synthesis of its temporality, its degree of freedom, and its relation to finitude.[29] The internal spiritual relation is how the self relates to itself, whether positive or negative.[30] Knots can form in the self when these relations become problematic. Kierkegaard designates an improper relation as a "negative unity."[31] For example, a negative unity forms when a person only conceives of their freedom in terms of how unfree they are. The self also has the capacity to adjust and affirm this condition. The relations composing the individual have to be engaged through the conditions of their actuality to become a positive unity.

There is also the external relation of the self. "The human self is such a derived, established relation, a relation that relates itself to itself and in relating itself to itself relates itself to another."[32] For Kierkegaard, the self as a mixture is established by god, and so the self is always also relating to god as well as to itself. But we can also understand this in relation to dependence on an external world in which we emerge. "This second formulation is specifically the expression for the complete dependence of the relation (of the self), the expression for the inability of the self to arrive at or to be in equilibrium and rest by itself, but only, in relating itself to itself, by relating itself to that which has established the entire relation."[33] A person does not establish their self, nor can they willfully decide their constitution or embeddedness in the world. Instead, humans exist in a relation of dependence with the entities and forces through and with which they exist. Establishing a proper relation to oneself requires coming to terms with this external relation as well.

Each side of the constitution of the self, internal and external, is linked to a different kind of despair. An improper internal relating in the self is "despair not to will to be oneself," or the desire to be a different self. Such a person wishes to have more freedom, less finitude, to be a different person. Improper external relating to the power that establishes the self is "despair to will to be oneself," or not wanting to be a self that is not under its own control. A person despairing to will to be oneself is caught in the disequilibrium of self-relating and cannot accept that they are not autonomous subjects but rather are established and maintained by something outside the self. In both cases, the problem of self-relating is inescapable. Existence "will nail him to himself so that his torment will still be that he cannot rid himself of his self."[34] Humans exist through a variety of possibilities and constraints but cannot escape either the fundamental terms of existence shape their self has taken in the world. To try to escape the despair of climate change is to try to escape existing as part a world tipping toward darkness. The problem of coming to terms with this even outside of the issue of climate change is so difficult that Kierkegaard argues that it brings everyone who confronts it to despair.

It is clear that for Kierkegaard, almost everyone is in despair, even those who focus on immediate materiality without reflecting on it, subject to only what seems to them to be good luck and bad luck. So, on the one hand climate change is not unique with regard to despair. On the other, climate change presses us to confront a condition that was always there but may not have been felt. Such a reckoning can help us confront what William E. Connolly calls passive nihilism, or the "formal acceptance of the fact of rapid climate change accompanied by a residual, nagging sense that the world ought not be organized so that capitalism is a destructive geologic force."[35] It is not surprising that it causes despair to deal with our place in a world organized in this way. For Kierkegaard, this is good. When such an event strikes someone, it can open reflection to despair, which is the first step forward. Climate change is both a particularly difficult and particularly pregnant occasion, given how it connects to both our external relations in the world and internal self-composition. It incites despair in the researchers who know it best and in activists whose efforts always come up short. Even the most wealthy and powerful find that despair about this event haunts them beyond whatever material fortifications they build.[36] And would it not be productive for someone living in material immediacy with a successful middle-class job, family, western lifestyle, etc. to fall into despair about that condition? Or even for those dreaming of the lifestyles promised by development to confront the fact that, in any case, it cannot be as promised? But the point is not to simply say that something is bad, or to resign oneself to a bleak future, or take on the weight of an immense guilt. Rather, the point is to take up the struggle of being both composed as we are and being reliant on the imperfect and messy world as we are.

This is Kierkegaard's point as well. Though one is always caught in the problem of self-relating, one is not always in despair. As Kierkegaard puts it: "The formula that describes the state of the self when despair is completely rooted out is this: in relating itself to itself and in willing to be itself, the self rests transparently in the power that established it."[37] This is also what Kierkegaard calls faith.[38] The problem is thus one of embracing existence on terms other than one's own. Thus, faith means willing one's own being as limited and established by forces beyond the self. It should be "transparently" clear in how a life is lived that we rely on a world and ecosystem beyond ourselves and that this ecosystem is the source of much of what we are. Ecosystem should here be understood as historical nature-culture relations of the kind that Haraway is so good at mapping. In this way, the individual is able to express the paradox and possibility that these larger forces could give existence to humans who, at the same time, are capable of imagining themselves and acting as separate, unique, and self-willed. Faith dissolves the autonomous individualism that is expressed in material acquisition and reason. The reason for existing is not understood and may not even exist but that is not necessary for accepting these conditions of existence.

Faith does not abandon the individual but develops it. One must "venture wholly to become oneself . . . Concern constitutes the relation to life, to the actuality of the personality."[39] Concern is the activity that actualizes a person. It constitutes an active attachment to life and the world. When we are concerned about something, we extend beyond ourselves and into that material and spiritual milieu which is outside of but also feeds into the self. Kierkegaard suggests a number of techniques for carrying out such extensions. He explains these techniques through the process of breathing. "Personhood is a synthesis of possibility and necessity.

Its continued existence is like breathing (*re*spiration), which is an inhaling and exhaling."[40] It is a matter of a movement in and out that expands beyond the self and returns to it. Reality is expanded without limitation by imagination according to different hopes, desires, fears, or intuitions, but then it is brought back into reconciliation with the actual form it could take within time. Haraway sees expanded possibilities and despair as exclusive of one another: "another world is not only urgently needed, it is possible, but not if we are ensorcelled in despair."[41] Yet Kierkegaard argues that the techniques for working through despair also enable us to think about alternative possibilities.

Techniques work on the key characteristics that constitute the self: temporality, freedom, and finitude. Infinitude risks carrying the self too far away from itself, detaching it from the world to live in extreme states of spirituality or abstraction. But too much finitude can make the self just a number, another individual among the mass pursuing vulgar and self-interested materialism. This latter condition can be countered by "volatilizing" feeling, knowing, and willing, propelling the self out toward the infinite. After being carried out a few times, one gets a sense of the ways that the self is not reducible to material self-interest. Then the right relation might become possible. "To become oneself is to become concrete . . . the progress of the becoming must be an infinite moving away from itself in the infinitizing of the self, and an infinite coming back to itself in the finitizing process. Every moment that a self exists, it is in a process of becoming, for the self . . . [in potentiality] does not actually exist, is simply that which ought to come into existence."[42] Both the finite and infinite are important for the self. If one is too materialistic, or too preoccupied with thought and spirit, practicing extensions in the other direction concretize the self. In this way, one becomes aware of the broader fields of potentiality that give form to the self.

Kierkegaard suggests similar experiments on the register of possibility and necessity, where the goal is to become oneself by achieving a movement in place.[43] One problem is that a person might be carried away by possibility, entirely losing actuality and instead using all of their energy pursuing the desires of the imagination. This is the paralysis that affects those for whom a bad version of hope has become a kind of climate denial, leaving them unable to grasp the troubling reality of the predicament.[44] But it is important to see what role necessity plays as well. "It takes time for each little possibility to become actuality . . . What is missing is essentially the power to obey, to submit to the necessity in one's life, to what may be called one's limitation."[45] Someone despairing over possibility is overly attached to a world or self that relies on a rapid or extensive realization of possibility. But by focusing on the relationship between possibility and time, for example, one in despair can rein this in to accept their existential limitations. Thus, movement in place considers many possibilities of a given situation but then returns to their existing situation to proceed modestly with one or more of them.

On the other hand, despairing over necessity may produce a philistine-bourgeois mentality, fatalism, or a determinism such that "everything has become necessary for a person or that everything has become trivial."[46] Interruptions in the normal order of things can spur such fatalistic views to recognize possibility. Such events, however, must "tear him out of [the miasma of probability] and teach him to hope and to fear" in such a way that his "being has been so shaken that he has become spirit by understanding that everything is possible."[47] In this case, mechanism is broken open to reveal one's spiritual self. This opens the spirit to belief in possibility. Activating the emotions of fear and hope to a degree directs thought and action beyond the constraints of necessity. Working through such possibility carries one beyond the understanding and thus requires belief to sustain them.[48] These techniques cannot produce faith on their own. They can only allow one to experience the paradox and possibility of existence.

It may be that Haraway pushes so hard against despair that she makes responding to a kind of determinism. "We must somehow make the relay, inherit the trouble, and reinvent the conditions for multispecies flourishing, not just in a time of ceaseless human wars and genocides, but in a time of human-propelled mass extinctions and multispecies genocides that sweep people and critters into the vortex. We must 'dare "to make" the relay; that is to create, to fabulate, in order not to despair.'"[49] She is right to fear despair because it is dangerous and potentially paralyzing and destructive, as Kierkegaard points out. But seemingly against the idea of staying with the trouble, her argument seems to be that in the face of catastrophe, part of the purpose of staying active is to avoid despair. Possibility fades in the drive for activity. But it may be that the shared genetic exchanges, ecosystem symbioses, and shared histories of objectification at the hands of biochemical industries are not yet sufficient to forging new multispecies relations. The success of new symbiotic projects is perhaps more dependent on what we don't know about a world that has so recently surprised us with its cultural-climatic feedback system, than the string figures that we consciously map and trace. Despair attunes to the paradox and absurdity of our existence in this world and at the same time open the door to possibility as belief, or intensification of passion, that motivates our responsiveness. Kierkegaard's techniques help us work through this process to gain the power of possibility without falling completely into despair.

Reflection is another technique for altering despair and responding to the event. Those who have developed capacities for reflection can gain distance, "so that despair, when it is present, is not merely a suffering, a succumbing to the external circumstance, but is to a certain degree self-activity, an act."[50] Reflection enables one to actively bring about or intensify a state of despair. Despair then begins to be seen as not simply an external event but as part of the constitution of the self. Reflection thus enables one to better respond to events. "He perceives that abandoning the self is a transaction, and thus he does not become apoplectic when the blow falls, as the immediate person does; reflection helps him to understand that there is much he can lose without losing the self. He makes concessions; he is able to do so."[51] Reflection enables one to actually begin to overcome despair in the event, since despair is now seen in the self and not just in the event. The ability to make concessions is the ability to give up part of the self and its attachments.

These are all techniques that help one engage an event so diffuse that it will not occur. They are productive because staying with the trouble is not an easy task. For many trying to respond to climate change, despair is part of that trouble.

Though Haraway marginalizes despair as a response, Kierkgaard suggests that it could be a necessary step in realizing our worldly condition. For him, "death is indeed the expression for the state of deepest spiritual wretchedness, and yet the cure is simply to die, *at afdøe*."[52] This seems contradictory, for death is precisely what is not possible. But what Kierkegaard means by *at afdøe* is more of a transition than an end. He sees the embrace of the terms of existence as a death that begins a more worldly life through faith. "The believer has the ever infallible antidote for despair—possibility . . . This is the good health of faith that resolves contradictions."[53] The cure for despair is to establish proper self-relating to the external. Imposing and seemingly limiting though they may be, the terms of existence are also the conditions for existence that make human life possible. One no longer despairs over being constituted by forces outside the self but rather seizes upon the possibility that allowed the self to come to be. The possibility for future becoming even in the face of climate change opens up. Yet Haraway rejects not just despair as an important response, but belief as well. The next section will clarify Kierkegaard's notion of the event and what believing in it might offer through an engagement with Haraway's resistance to it.

The unequal occasion: committing to the event

So far, I have discussed two of Kierkegaard's approaches to the event. The previous chapter looked at a poetic approach that emphasizes its uncertainty and possibility as well as way to attract us to attempts to respond to it, despite the anxiety involved in doing so. Alternatively, amidst the trouble of a diffuse event, the problem of despair can be worked through using a variety of techniques that modify the way the self relates to the world in order to build responsiveness. In *Philosophical Fragments* Kierkegaard gives his most detailed exposition of the event. I will explicate nine characteristics of the event that clarify and build on the previous two approaches. The goal is to better understand how belief can meet the urgency of climate change.

Before getting to Kierkegaard, however, it is necessary to address an issue that has been simmering in background so far: Kierkegaard's embrace of belief and Haraway's opposition to it. "The refusal of the modernist category of belief is also crucial to my effort to persuade us to take up the Chthulucene and its tentacular tasks."[54] Haraway does not fully explain the problem with belief but does point to a couple key problems. First, "believing is not sensible."[55] By this she means that it is not material, or about practices of worlding. For her, believing in climate change is about religion and confession and thus opposed to the sciences. Second, "beliefs and commitments are too deep to allow rethinking and refeeling."[56] Haraway is right to resist both dogmatism and transcendence but perhaps moves to quickly to dismiss belief and the broader theological mode. This may not be so surprising given the "case of permanent raging indigestion" she still harbors from her early experiences in the church.[57] Nonetheless, as I have been trying to argue, Kierkegaard's approach to belief is sensible, worldly, and so difficult that it never becomes an inflexible commitment but remains a striving and experimenting. For these reasons, I think it can be helpful for her task of staying with the trouble. While the last section spoke of the importance of despair in connection to climate change, Catherine Keller, whose work will be explored in more detail in Chapter 4, tries to bring Haraway's work into connection with theological approaches to time and hope.[58] This section develops belief as a sensitivity to the uneven process by which novel events come into existence and a commitment that reshapes the believer's relations to the world.

Kierkegaard outlines his theory of the event by contrasting it with that of Socrates. For Socrates, "every human being is himself the midpoint, and the whole world focuses only on him because his self-knowledge is God-knowledge . . . the temporal point of departure is a nothing, because in the same moment I discover that I have known the truth from eternity without knowing it."[59] In this view, everybody already knows the truth even if they do not recognize it. Every moment and encounter is a potential occasion for recalling the truth, which also means that no particular moment or encounter is required to learn the truth. The student owes nothing essential to the teacher. Though the teacher may have been the occasion, the occasion could just as easily have been something else. Though Kierkegaard attributes this to Socrates, we might read it more broadly as the abstract Enlightenment view of the world, in which knowledge of the world is complete and the particular material time and place of the individual are accidental and secondary. Some with this view become climate change denialists, seeing it as just another occurrence in the universe that will continue largely unaffected by what happens on this planet.

For Kierkegaard, the particular occasion of the event engenders a different relation to the truth than all other occasions. He argues that people are unable to understand the truth and are even unaware of this condition. Both the truth and the condition for learning it require the event. "If the learner is to obtain the truth, the teacher must bring it to him, but not only that. Along with it, he must provide him with the condition for understanding it . . . because the condition for understanding the truth is like being able to ask about it—the condition and the question contain the conditioned and the answer."[60] The "teacher" expresses how the moment intervenes into existence. Kierkegaard often calls the event "the moment" to emphasize the novelty and interruptive force of its emergence. The importance of the event as Kierkegaard conceives it is that it does not just disturb knowledge and habit but fundamentally changes the condition of being, and thus what knowledge is possible and what will count as knowledge. The event disrupts those experiences, intuitions, understandings, and perceptions that lead to the formation of questions about truth. In so doing, it creates a condition for asking questions that previously were not available.

In this way, the first characteristic of the event is that it produces existential transformation.

> Inasmuch as [the learner] was in untruth and now along with the condition receives the truth, a change takes place in him like the change from 'not to be' to 'to be,' but this transition from 'not to be' to 'to be' is indeed the

transition . . . by which he enters the world a second time just as at birth—an individual human being who as yet knows nothing about the world into which he is born, whether it is inhabited, whether there are other human beings in it.[61]

This rebirth is not to a transcendent domain but deeper into the world, which previously had been experienced superficially. The "truth" is not a matter of knowledge but of one's existential coordinates in the sense that it reshapes our sense of self and place. In such a situation, it is necessary to begin again to develop a sense of awareness about the world: to begin to think and investigate freshly about what exists, how it hangs together, with what other beings it does so, what matters, what forms of living correspond to it, and what truth is.

The second characteristic of the event is that it is a sensual approach to truth, in contrast to an abstract mental one. "Whereas the Greek pathos focuses on recollection, the pathos of our project focuses on the moment, and no wonder, for is it not an exceedingly pathos-filled matter to come into existence from the state of 'not to be'?"[62] The event carries an experience, a feeling. For Kierkegaard, it requires sensitivity and passion. Beyond propelling a rethinking of what it is to come into being, the event heightens our sensitivity to the process of coming into existence. Belief is the name Kierkegaard gives to this pathos: "Belief is a sense for coming into existence."[63] Belief is not a commitment to the factual knowledge of the event. Rather, in approaching it through passion, calls the fact of the event into question.[64] It resolves to sensually commit to its existence without nailing it to the static form of a fact. Thought can only approach the edge of the event, beyond which it breaks down.

Thought cannot proceed further because of the third characteristic, which is that the interruption of the event forms an "unequal occasion."[65] For Kierkegaard, disequilibrium structures the event because god, who is motivated by love, intervenes in the world to teach the truth. The nature of this disequilibrium is that the event does not need humans but they depend on it.[66] Bruno Latour deploys a different but analogous framing of climate change in terms of Gaia: "She is at once extraordinarily *sensitive* to our action and at the same time She follows goals which *do not* aim for our well-being . . . She will last . . . We are the ones who are in trouble."[67] The problem is how to overcome this inequality and bridge this existential gap. Kierkegaard argues that this is a problem for both god and humans, though each has a different solution. God's solution is to manifest as a human servant who must endure everything exactly as a human would. The difference between god and the learner is made equal through god's resolution to become human. We might see a similar move in the way Latour refers to the breakdown of the nature/culture divide in the Anthropocene: "Gaia-in-us or us-in-Gaia."[68] Climate change signals nature becoming human in a new way.

But the learner must also have the resolution to believe in god's human transformation, to believe in nature's new intimate link with humanity. Kierkegaard explains belief through the interaction between thought and the unknown. Thought tends toward its own downfall in its attraction to that which it cannot understand.

"This, then, is the ultimate paradox of thought: to want to discover something that thought itself cannot think. This passion of thought is fundamentally present everywhere in thought, also in the single individual's thought insofar as he, thinking, is not merely himself. But because of habit we do not discover this."[69] When we think, we are not wholly ourselves. Thought, in its innate relation to that which it does not understand, draws us toward the unknown. Yet much of the time we do not notice this because of the life-guiding habits and assumptions that occupy thought. By desiring to discover what it does not know, thought gives the unknown a permanent and potentially destabilizing position within the self.

Once we engage the unknown, thought begins to call itself into question.[70] Haraway draws on Isabelle Stengers to make this point, as "Gaia is an intrusive event that undoes thinking as usual."[71] This destabilization of the thought begins the transformation that will produce a rebirth of the self. "But what is this unknown against which the understanding in its paradoxical passion collides and which even disturbs man and his self-knowledge? . . . let us call this unknown *the god*. It is only a name we give to it."[72] Kierkegaard is not only making a religious argument, though that aspect is crucial for him. The learner struggles with the unknown in the same way that the learner struggles with god's intervention into the world: they are both unassimilable to the identity or understanding of the self-sufficient subject. In both cases, the learner is confronted with disequilibrium. When the individual encounters the disequilibrium of the event, they recognize their condition as that of untruth, as being unable to understand what is happening. The discomfort that this causes is the fourth characteristic of the event. "Through the moment, the learner becomes untruth; the person who knew himself becomes confused about himself and instead of self-knowledge he acquires the consciousness of [offense] . . . all offense is in its essence a misunderstanding of *the moment*."[73] This is the discomfort that the existential disequilibrium of the event creates. Kierkegaard calls this feeling of being lost, overwhelmed, and excluded from a central part of reality offense. This discomfort makes the urgency of the moment felt: it presses upon us and to rid ourselves of it we either try to turn away or try to embrace it.

Given this discomfort, is it actually so surprising that people have by and large failed to accept the reality of climate change? Indeed, many of the forms of denial discussed in the introduction result from the attempt to avoid this discomfort of misunderstanding and being lost in the world. From insistence on the powers of reason to understand and control the environment, to psychological instincts that direct thought away from disturbing threats, to the reliance on the stability of cultural norms to provide emotional security. Kierkegaard would see each of these as symptomatic of offense. Climate change presents a novel situation that is still not well understood. Further, it fundamentally calls into question the place of humans in the world and opens new questions about the terms of existence. Climate change may be an event whose reality we need to commit to without being able to understand it, even though it makes us feel uncomfortable and insecure.

According to Kierkegaard, the individual remains in the discomfort of offense as long as they continue to try to understand the event. "Precisely because offense is a suffering in this manner, the discovery, if it may be put this way, does not belong to the understanding but to the paradox, for just as truth is *index sui et falsi* [the criterion of itself and of the false], so also is the paradox, and offense does not understand itself but is understood by the paradox."[74] As long as one takes the suffering of offense as the problem, then one will remain within offense. One might, for example, be resentful of a world that is ripe for human flourishing but that also contains many forces stronger than humanity and can thus become quite inhospitable to it. The key is to shift to the view of the event (paradox), but this shift cannot be accomplished by the understanding since understanding cannot comprehend the event. From within offense, the understanding declares that "the moment is foolishness, the paradox is foolishness—which is the paradox's claim that the understanding is the absurd but which now resounds as an echo from the offense."[75] The understanding claims that the paradox is absurd, ridiculous, extreme, or too different and therefore refuses to deal with it. But the paradox of the event transparently declares that both itself and the understanding are absurd. The existence of climate change is contingent, just as the existence of humans is. This is why Kierkegaard thinks of offense as an "acoustical illusion," because the understanding's declaration that the paradox is absurd is really an echo, a parroting of what the paradox itself declares. Thus, the fact that the paradox is the criterion of both itself and the offense. To come to terms with the event, one must leave the understanding and accept the paradox.

The fifth element is the way belief bridges the disequilibrium of the paradox. From the side the event, god descends to the level of man or nature becomes human in the Anthropocene. From the side of the human, the problem is to overcome offense not by understanding the event but also by understanding the difference between the human and the event. The Anthropocene does not make humans and nature the same. To believe is to "understand" through passion and feeling that something different comes into being. Belief is the pathos that accepts the paradox, the difference, inherent in the break of the event. "We do not say that he is supposed to understand the paradox but is only to understand that this is the paradox. . . [that] in which this occurs . . . is that happy passion to which we shall now give a name, although for us it is not a matter of the name. We shall call it *faith*."[76] If nature becomes human, then the understanding as the principle of identity must also be relaxed to accept the difference of the paradox. Belief is the passion to leave behind one's established self and embrace the transformation entailed by something new coming into existence. We could say that if we believed in climate change, we would not understand it, but we might "come to an understanding" with it and the terms of existence it presents. This belief would entail a kind of rebirth with a new way of existing in the world. For its part in this happy meeting, the world might then continue to support human flourishing.

The sixth characteristic is that one does not have to experience the event in its immediacy to believe in it and undergo its change. This is not a refusal of sensual experience, which Kierkegaard puts beyond doubt. Rather, the illusiveness of the event is that its process of coming into existence cannot be immediately experienced.[77] It is in this illusiveness and uncertainty that belief forms its bonds.

"The same is true of an event. The occurrence can be known immediately but not that is has occurred, not even that it is in the process of occurring, even though it is taking place, as they say, right in front of one's nose . . . and therein lies the transition from nothing, from non-being, and from the multiple possible 'how.'"[78] While immediate experience knows with certitude what exists, it does have any sense for what it would mean to not take immediacy as fact or for the process of coming into existence.

The seventh characteristic of the event is that it only becomes a historical fact through faith.[79] According to Kierkegaard, one of Christ's contemporaries who saw him regularly would not thereby have any way of recognizing him as god. He suggests a number of ways that people might try to get the most accurate account of the event, pointing out that none of this would bring one closer to believing in it. Rather, all that later generations need to hear is that their predecessors had believed that such and such an event occurred. The report unsettles immediacy and comes into existence in the same way as the event. It is the same today with regard to climate change and the attempt to date the beginning of the Anthropocene. This means that neither someone present at the invention of the steam engine, nor the trinity test, nor a climate scientist taking atmospheric measurements, nor a farmer experiencing rapid changes in local conditions are closer to knowing the event of climate change. While identifying a date may be important for the practice of geology, it is not important for accepting the emergence of anthropogenic climate change on Earth. The dispute over when to place it attests to the insufficiency of any particular sensual experience and the need to embrace its occurrence through such uncertainties and possibilities. Belief, while constantly maintaining a relation to these uncertainties, carries with it the passion that commits it to the coming into existence of the event.

The eighth characteristic of the event is what Kierkegaard calls eminent faith. "Faith must be taken in the wholly eminent sense, such that this word can appear but once, that is, many times but in only one relationship."[80] While one hand faith establishes a concrete historical moment, on the other it encompasses the unity of every instance of faith. It connects individuals through the relation they actively adopt toward the event. This serves an interesting function for a dispersed event like climate change: dispersed in time, geography, cause, effect, justice, awareness, etc. Kierkegaard's point is not that it erases this unevenness and dispersion but that believers become connected through expressing the same relation toward the process of coming into existence. Their approach is to be able to break out of a progressive-teleological view and "pause (here wonder [belief] stands *in pausa* and waits for the coming into existence), which is the pause of possibility."[81] This goes beyond a notion of the event attached to a particular religious tradition, and even beyond any specific event. What Kierkegaard seems to favor is a general sensibility for and responsiveness to the process of coming into being. This highlights the new relation to the world established in the rebirth of the individual through belief.

The ninth characteristic is that the connection that faith establishes between all individual instances of the event can bring about a different relation to time.

"But for those who are very different with respect to time, this latter equality absorbs the differences among those who are temporally different . . . Every time the believer makes this fact the object of faith, makes it historical for himself, he repeats the dialectical qualifications of coming into existence."[82] There is a sense in which time loses its consistency and becomes instead an expression of a singular event with multiple occurrences. On one hand, this is the other side of the view that every moment could potentially be an event, except that now every event marks a moment of time. On the other hand, this goes beyond the sense of time as an entity that plays out in increments. If time can be subsumed under the event, then what results on the side of the individual is something like the atemporal condition (nonbeing) out of which the event emerged. Thus, Kierkegaard connects the faith that unifies all events with an ability to repeat the process of coming into being. What he suggests is that the believer can continue to undergo the process of rebirth, perhaps even becoming an event in doing so.

Believing in the event may seem an esoteric and useless endeavor. But it is also possible that when we consider our experience in light of what Kierkegaard has to say, there may be something to it. On the one hand, it delegitimizes those who do not believe in climate change on the basis of a dogmatic and fundamentalist Christianity. While Kierkegaard's notion of belief is forceful, it is neither dogmatically grounded in scripture nor fundamentalist. Rather, it is an existential endeavor grounded in uncertainty, anxiety, and distress. Furthermore, it entails a robust worldliness that is the mark of a faith not content to make this life second to an afterlife. Yet this notion of belief also cuts into the facile binary in the United States today between those who reject climate change on religious grounds and those who believe in it on the basis of science. At the heart of this apparent divide is the fact that both those who accept and those who deny climate change live similar climatically destructive lives. In the light of Kierkegaard's thought, Christian fundamentalism may seem more like an angry outburst rooted more in secular-capitalist lifestyles than Christianity. At the same time, acknowledgment of climate change based on information comes all too easily to rationalized subjects who understand climate change without allowing it to touch their lives.

Believing in climate change entails an impassioned staying with the trouble of the event as it casts human existence as untruth. Belief would not need precise information about climate change, nor does it need to try to predict exactly how it will occur. Rather, just the report that climate change is occurring is enough to incite the uncertainty and anxiety that can serve as the foundation for a response to climate change. The existential change brought on by belief would manifest throughout our lives, not as a specific issue to which we give part of our conscious attention. Finally, a believer would love the world that could produce such a dramatic shift, realizing that the conditions that established the believer's own life are as paradoxical and beyond control as those that made climate change possible.

***

In the months before his death, Kierkegaard made himself into an event. In addition to a slew of newspaper articles, he put out nine issues—the tenth was written, but not published until after his death—of his own broadsheet entitled *The Moment*. These writings were an all-out attack on the State Church. He did this because he felt that it was necessary to give a report to his contemporaries.[83] He felt that this was necessary in part because his society had no experience of the event.[84] In these writings, he takes many polemical positions such as: Christians would be better off if they stopped attending church; the Christianity of the New Testament does not exist; "That the Pastors Are Cannibals, and in the Most Abominable Way;" and that atheists are better than pastors. There is a sharp edge to this thought and writing that cuts into routinized and easy Christianity. Kierkegaard describes his task in this way:

> The point of view I have set forth and do set forth is of such a distinctive nature that I quite literally have no analogy to cite, nothing corresponding in eighteen hundred years of Christianity. In this way, too—facing eighteen hundred years—I stand quite literally alone. The only analogy I have before me is Socrates; my task is a Socratic task, to audit the definition of what it is to be a Christian—I do not call myself a Christian (keeping the ideal free), but I can make it manifest that the others are that even less.[85]

Even here, in this critical mode, belief is not the easily attained dictatorship of dogma. Rather, he emphasizes it as a difficult task that he has not completed but that has not even been recognized by those who claim to have done so. To do so, he highlights a number of essential Christian principles in relation to improper modes of "belief" perpetuated under the name of Christianity. In this way, he becomes an event with the potential to cause a break in the existing order. He puts everyone to the decision to change themselves.

Though many things divide Kierkegaard and Pope Francis, they may share an interruptive potential. Francis' encyclical on climate change "Laudato Si'" disrupted the vision that many non-Catholics had of the Church as well as that of many Catholics within it. Bill Maher, who generally has nothing positive to say about religious people anywhere, felt compelled to laud Francis' engagement on an issue that, in his understanding, fell beyond the bounds of religion.[86] Indeed, the encyclical may signal a conversion experience for the Church itself that could overturn its existing history and theology.[87] Certainly, Francis has tapped into and helped spread the disruptive nature of this event. Laudato Si' as a report on climate change may serve as an interesting comparison to the IPCC reports, which despite their increasingly dire predictions, are not as unsettling and striking as the former. It also seems that Laudato Si' professes and enacts a greater commitment to the reality of climate change than the cool and casual comments about it by well-educated liberals like Maher, who profess environmental concerns as their territory. And yet an engagement with Laudato Si' goes further. In it, Francis addresses not just members of the Catholic Church, but "every living person on this planet."[88] Amitav Ghosh recognizes in this call and those of other religious leaders to respond to climate change a challenge to the presumption that the nation-state system is the best form of organization we have so far for responding.

Indeed, the transnational, long-term, anti-economistic, and capacity to think about catastrophe (the event) are all practical tools that these social organizations have at their disposal and that nation-states so far seem to lack.[89] Such a material benefit might even be able to convince Haraway of the benefit of these systems of belief.

Laudato Si' intensifies the climate event in significant ways. Kierkegaard often equated faith with a kind of madness, which is how many treat Francis's critiques of the capitalist economic system, the "technocratic paradigm," and "throwaway culture."[90] But he goes even further in his calls to treat Earth like a mother, and other organisms as siblings. How should one understand such a suggestion, take it seriously, and live by it? The science is important to him and receives some discussion in the encyclical. Yet for him it is not a matter of trying to understand the issue but rather of becoming "painfully aware" of it.[91] Indeed, he sees the Earth as composed of an integral ecology that can only be partially grasped through the understanding but which also requires aesthetic sensibility, a sense of meaning and purpose, and perhaps belief as well.[92] Thus, Francis calls for conversion to an "ecological spirituality" that "can motivate us to a more passionate concern for the protection of our world."[93] This sensitive, passionate approach to climate change seems to embody the pathos that Kierkegaard calls belief. In this way, Francis seeks to interrupt our contemporary mode of being, to audit it, and bring us a report of the climate event, hoping to push us toward belief in it and the existential transformation it entails.

Notes

1 Marshall, *Don't Even Think About It*, 225.
2 Ibid., 229.
3 Connolly, *Facing the Planetary*, prelude.
4 Ibid., 157.
5 Throughout this project I will refer to Kierkegaard rather than his pseudonyms. This is not because they are irrelevant, but because I read in Kierkegaard what Deleuze and Guattari call the multiplication of the self in *A Thousand Plateaus*. Given the problem of writing as a multiple self, I can agree with Alastair Hannay when he argues that "pseudonymity can just as well be an effective means of exposure, the disguise of a disguise that allows an author to spill more of himself onto his pages than would be prudent or proper if the works were signed" (Hannay, *Kierkegaard*, x). Prudence is not just a matter of social standing, but of philosophical and authorial coherence and legitimacy. Through his pseudonyms, Kierkegaard was able to explore and dramatize a greater variety of thoughts and experiences than one might normally be allowed while also being taken seriously.
6 Haraway, *Staying with the Trouble*.
7 Ibid., 1.
8 Ibid.
9 Ibid., 12.
10 Ibid., 36.
11 Ibid., 4.
12 Miéville, "The Limits of Utopia."
13 Ibid. Haraway would resist Miéville making hope into a weapon and the combative version of struggle that implies. I will address this in chapter four, in the context of Latour's appropriation of Schmitt.

14 Harayway, *Staying with the Trouble*, 3–4.
15 Ibid.
16 Salamon, "What S-Town Gets Wrong."
17 Norgaard, *Living in Denial*.
18 Kierkegaard, *The Sickness unto Death*, 17–18.
19 Haraway, *Staying with the Trouble*, 3.
20 This position can be contrasted with that of James Lovelock, who argues that we should not try to mediate climate change and simply enjoy ourselves while fortifying human life in the most efficient way possible. See Lovelock, *A Rough Ride to the Future*; Aitkenhead, "James Lovelock."
21 Kierkegaard, *Sickness unto Death*, 7.
22 Ibid., 18.
23 Franzen, "What If We Stopped Pretending."
24 Hamilton, *Requiem for a Species*, 216–17.
25 Kierkegaard, *Sickness unto Death*, 18.
26 Ibid., 18–19.
27 Ibid., 18.
28 Ibid., 20.
29 Ibid., 13.
30 Ibid.
31 Ibid.
32 Ibid., 13–14.
33 Ibid., 14.
34 Ibid., 21.
35 Connolly, *Facing the Planetary*, 9.
36 See Osnos, "Doomsday Prep for the Super-Rich"; Karlis, "We Asked Psychologists Why So Many Rich People Think the Apocalypse Is Coming."
37 Kierkegaard, *Sickness unto Death*, 14.
38 Ibid., 82.
39 Ibid., 5.
40 Ibid., 40.
41 Haraway, *Staying with the Trouble*, 51.
42 Kierkegaard, *Sickness unto Death*, 30. Translators' brackets.
43 Ibid., 36.
44 Hamilton, *Requiem for a Species*, 130–3.
45 Kierkegaard, *Sickness unto Death*, 36.
46 Ibid., 40.
47 Ibid., 40–1.
48 Ibid., 38.
49 Haraway, *Staying with the Trouble*, 130.
50 Kierkegaard, *Sickness unto Death*, 54–5.
51 Ibid.
52 Ibid., 6. I have left *at afdøe* in Danish since it has a more complex meaning than the rendering given by Hong and Hong as "to die to the world." While *at døe* is literal physical death, *at afdøe* is a sort of abstract death that can mean leaving a way of life or state of being to enter a new one. Here, transformation and continuation are more important than the sense of an end. Indeed, *at afdøe* can even imply the idea of a new life. I am indebted to Johanna Treider for this clarification.
53 Kierkegaard, *Sickness unto Death*, 39–40.
54 Haraway, *Staying with the Trouble*, 42.
55 Ibid., 88.
56 Ibid., 208.
57 Ibid., 179.

58 Keller, *Political Theology of the Earth*, 87–90.
59 Kierkegaard, *Philosophical Fragments*, 11–13.
60 Ibid., 14.
61 Ibid., 19.
62 Ibid., 21.
63 Ibid., 84.
64 Ibid.
65 Ibid., 25.
66 Ibid., 24–5.
67 Latour, "Waiting for Gaia."
68 Ibid.
69 Kierkegaard, *Philosophical Fragments*, 37.
70 Ibid., 39.
71 Haraway, *Staying with the Trouble*, 44.
72 Kierkegaard, *Philosophical Fragments*, 39.
73 Ibid., 51.
74 Ibid., 50. Editors' Brackets.
75 Ibid., 52.
76 Ibid., 59.
77 Ibid., 81.
78 Ibid., 81–2.
79 Ibid., 87.
80 Ibid.
81 Ibid., 80. First set of brackets mine, second set editors'.
82 Ibid., 88. My brackets.
83 Kierkegaard, *The Moment and Late Writings*, 73–4.
84 Ibid., 129–37.
85 Ibid., 340–1.
86 Maher, "Real Time with Bill Maher: Pope Frank in America."
87 Klein, "A Radical Vatican?"
88 Francis, "Laudato Si'."
89 Ghosh, *The Great Derangement*, 159–62.
90 Francis, "Laudato Si'."
91 Ibid.
92 Ibid.
93 Ibid.

Bibliography

Aitkenhead, Decca. "James Lovelock: 'Enjoy Life While You Can: In 20 Years Global Warming Will Hit the Fan'." *The Guardian*, March 1, 2008. www.theguardian.com/the guardian/2008/mar/01/scienceofclimatechange.climatechange.

Connolly, William E. *Facing the Planetary: Entangled Humanism and the Politics of Swarming*. Durham: Duke University Press, 2017.

Deleuze, Gilles, and Félix Guattari. *A Thousand Plateaus: Capitalism and Schizophrenia*. Translated by Brian Massumi. Minneapolis: University of Minnesota Press, 1987.

Franzen, Jonathan. "What If We Stopped Pretending." *The New Yorker*, September 8, 2019. www.newyorker.com/culture/cultural-comment/what-if-we-stopped-pretending.

Ghosh, Amitav. *The Great Derangement: Climate Change and the Unthinkable*. Chicago: University of Chicago Press, 2016.

Hamilton, Clive. *Requiem for a Species*. New York: Earthscan, 2010.

Hannay, Alastair. *Kierkegaard: A Biography*. New York City: Cambridge University Press, 2001.

Haraway, Donna. *Staying with the Trouble: Making Kin in the Chthulucene*. Durham: Duke University Press, 2016.

Karlis, Nicole. "We Asked Psychologists Why So Many Rich People Think the Apocalypse Is Coming." *Salon*, July 16, 2018. www.salon.com/2018/07/16/we-asked-psychologists-why-so-many-rich-people-think-the-apocalypse-is-coming/.

Keller, Catherine. *Political Theology of the Earth: Our Planetary Emergency and the Struggle for a New Public*. New York: Columbia University Press, 2018.

Kierkegaard, Søren. *The Moment and Late Writings*. Edited and Translated by Howard V. Hong and Edna H. Hong. Princeton, NJ: Princeton University Press, 1998.

Kierkegaard, Søren. *Philosophical Fragments*. Edited and Translated by Howard V. Hong and Edna H. Hong. Princeton, NJ: Princeton University Press, 1985.

Kierkegaard, Søren. *The Sickness unto Death*. Edited and Translated by Howard V. Hong and Edna H. Hong. Princeton, NJ: Princeton University Press, 1980.

Klein, Naomi. "A Radical Vatican?" *The New Yorker*, July 10, 2015. www.newyorker.com/news/news-desk/a-visit-to-the-vatican.

Latour, Bruno. "Waiting for Gaia. Composing the Common World Through Art and Politics." In *French Institute Lecture*. London, November 2011. www.bruno-latour.fr/sites/default/files/124-GAIA-LONDON-SPEAP_0.pdf.

Lovelock, James. *A Rough Ride to the Future*. New York: The Overlook Press, 2014.

Maher, Bill. "Real Time with Bill Maher: Pope Frank in America." *HBO*, September 25, 2015. www.youtube.com/watch?v=C1EXlQ7iLZU.

Marshall, George. *Don't Even Think About It: Why Our Brains Are Wired to Ignore Climate Change*. New York: Bloomsbury, 2015.

Miéville, China. "The Limits of Utopia." *Salvage*, no. 1 (2015): 177–89.

Norgaard, Kari Marie. *Living in Denial: Climate Change, Emotions, and Everyday Life*. Cambridge: The MIT Press, 2011.

Osnos, Evan. "Doomsday Prep for the Super-Rich." *New Yorker*, January 30, 2017. www.newyorker.com/magazine/2017/01/30/doomsday-prep-for-the-super-rich.

Pope Francis. "Encyclical Letter Laudato Si' of the Holy Father Francis on Care for Our Common Home." *The Holy See*, May 24, 2015. http://w2.vatican.va/content/francesco/en/encyclicals/documents/papa-francesco_20150524_enciclica-laudato-si.html.

Salamon, Margaret Klein. "What S-Town Gets Wrong About Climate and Mental Health." *The Climate Psychologist*, May 5, 2017. http://theclimatepsychologist.com/what-s-town-gets-wrong-about-climate-and-mental-health/.

3   The self in the social
Sensitivity to an eventful world

Who is the subject of climate change?

The previous two chapters looked at the problem of liberal justice responses to climate change and suggested Kierkegaard's thought as one way to begin moving beyond them. Yet his form of commitment can be understood as problematically individualizing. Though he grounds his form of commitment in the insufficiency of the self as determined by forces outside it, that self is still treated relatively independently of social relations. This chapter will situate the event and the self in a deeper social context. The best place to start such an inquiry is in the debate of how to name to era that we are in. The Anthropocene and its alternatives such as Capitalocene are as much about identifying the nature of the problem and who is responsible, as they are about marking the beginning of this era.

Not everyone is familiar with the term Anthropocene, yet it has become the primary way of designating the era of climate change for journalists and academics.[1] Nonetheless, there is significant and well-reasoned opposition to this term. Perhaps the most thorough analysis and critique comes from Christophe Bonneuil and Jean-Baptiste Fressoz. To abbreviate an extended analysis, they make four main points. First, that the term Anthropocene and the way it is analytically used totalizes human actions under one collective biological entity that does not really exist because it is abstracted from social relations and inequalities.[2] Second, that it presents itself as an ecological awakening while covering up a long history of critique of ecological destruction.[3] Third, in so doing, it establishes a divide between scientific experts who understand the problem and can determine the response and a docile lay public who follows.[4] Fourth, this becomes basis for a new form of control called "geopower" that uses the Earth system as a site of calculation and control.[5] To sum up, "the subject of the Anthropocene, moreover, appears as an eco-citizen optimizing her carbon credits, managing her individual footprint (and governed by way of her environmental reflexivity). This is a being plugged into the flows of 'ecosystemic services' that the different compartments of the Earth system supply her with. Thus, the subject of the Anthropocene is constructed as a passive public that leaves solutions to geocratic experts."[6] This, then, is the pragmatic texture of the program for which the liberal justice image of thought explored in Chapter 1 serves as the moral and ideological justification.

In both cases, social differences are erased and substituted by functionally similar economic individuals. The subject of the Anthropocene thus appears collectively homogenous and individually undifferentiated.

While this analysis makes any uncritical use of the term Anthropocene problematic, some have gone to lengths to try to critically refashion the term while drawing on the momentum it has developed. Bonneuil and Fressoz, somewhat unjustly, lump Bruno Latour and Dipesh Chakrabarty in with the rest of the Anthropocenologists. In doing so, they overlook the extent to which the former uses it not to empower science but to pursue precisely the kind of environmental humanities program they favor while also attending to the need to compose an as yet nonexistent subject of the Anthropocene on the basis of existing differences.[7] In the case of the latter, they overlook the extent to which Chakrabarty starts with the existence of social difference but then moves to investigate sources of commonality at this particular historical juncture.[8] To these figures would have to be added William E. Connolly's hesitant use of the term given the problems associated in order to bring the natural sciences and humanities into conversation and speak to the condition that he calls "entangled humanism."[9] These thinkers all push the term in other than simply empowering scientists, re-inscribing liberal subjects, and using the ecology as a new framework for domination.

Finally, one would have to look to some of the alternatives. Bonneuil and Fressoz proliferate labels: Phagocene, Polemocene, Thanatocene, and Agnotocene, just to name a few. Perhaps to two most prominent alternatives, however, are the Capitalocene and the Chthulucene. The former has been used by a number of thinkers but was popularized by Jason W. Moore who places the beginning of our era after 1450 to "prioritize the relations of power, capital, and nature that rendered fossil capitalism so deadly in the first place."[10] These relations, he argues, are the source of the problem and what needs to be reformed. Haraway favors using multiple terms. Anthropocene and Capitalocene have their advantages and disadvantages but thinks that we need to add Chthulucene as a way of moving forward. On one hand, the Chthulucene is "an ongoing temporality that resists figuration and dating and demands myriad names" and, on the other, "is made up of ongoing multispecies stories and practices of becoming-with . . . in precarious times."[11] Thus, she pushes us toward the question of how particular multispecies assemblages can repattern the pattern they have been handed for the purposes of ongoing collective survival.

I follow Haraway in thinking that all the terms have their use and that there is no need to be exclusivist. Each term helps call attention to a particular angle of the problem and none of them can catch them all. Terms such as Capitalocene and Oliganthropocene that emphasize inequalities in carbon emissions and impacts suffered have particular importance right now. And Latour is right in suggesting that a political agent (or collection of agents) capable of responding adequately to climate change does not yet exist. Doing so requires putting the different scales of climate change into connection. Félix Guattari pointed in this direction when he developed ecosophy as "an ethico-political articulation . . . between the three ecological registers (the environment, social relations and human subjectivity."[12]

Looking at environmental transformation in connection to social relations and subjectivity shows both the obstacles impeding responsiveness and the sources of more efficacious agency.

This chapter will develop a Nietzschean ethos of the event to engage climate change on these three registers. First I will look at how Nietzsche uses seasons, times of day, and festivals to both alert us to the way events recur within us, interrupting the stable periods of subjective life, and to characterize affirmative and decadent dispositions toward these shifts. A discussion of Klossowski's reading of the Eternal Return carries us to Nietzsche's focus on drives and experiments that rework the self through the event. The significance of this dimension of the event is discussed in relation to the denial of climate change at the emotional level as well as the potential for unexpected emotional eruptions to forge stronger responses to climate change. This aspect of his thought informs a politics of experimentation as a way to change the habits, dispositions, and modes of being that help propel climate change today.[13] Finally, I connect two of Nietzsche's temporally diffuse events—that of the human and the death of God—to examine the place of humans in the world. These events attune us to the problem of finitude to avoid resenting a world that is not predisposed to human survival.

Seasons of the subject

Nietzsche loves the change of seasons and its energy infuses his work. *The Gay Science* expresses the spirit of spring. "It seems to be written in the language of the wind that brings a thaw: it contains high spirits, unrest, contradiction, and April weather, so that one is constantly reminded of winter's nearness as well as of the *triumph* over winter that is coming, must come, perhaps has already come . . . Gratitude flows forth incessantly."[14] This elevation of the seasons alerts us to dynamics of experience to which we often fail to attend. Spring is a season, a health, a disposition, a change, a sudden and strong infusion, and an event. We cannot locate the change it brings exactly and it catches us by surprise: we find ourselves caught up in the uncertain energy of the transition as it wells up within us, infusing experience with a different outlook. Thus Nietzsche draws on a common experience of subjective and psychological transition to positively valuate the idea of interrupting an established routine to begin something new. This is how he acquaints us with what he calls the "art of transfiguration."[15]

Nietzsche draws on other transitions that recur like the seasons to show how the interruptive moments of life trade off with settled periods. Evening and daybreak capture the same spiritual disposition and movement of energy as seasonal transitions.[16] Here, he clarifies that it is not just a matter of the repeated surprise of the transition but of the dangerous temptations of evening.

> It is not wise to let the evening judge the day: for it means all too often that weariness sits in judgment on strength, success and good will. And great caution is likewise in order with regard to *age* and its judgment of life, especially as, like evening, age loves to dress itself in a new and enticing morality and

knows how to put the day to shame through twilight and solemn or passionate silence . . . from now on he wants to found, not structures of thought, but institutions which will bear his name . . . he will invent a religion . . . a binding institution for future mankind.[17]

The evening disposition tends toward comfort and routine, having lost its creative power, its willingness to attempt new ways of living. Not only does such a disposition forgo the novelty of daybreak but it also seeks to withhold it from others as well. The evening disposition justifies its lassitude by constructing institutions, limits, and even a religion to control the moralities and existential options available to others. Nietzsche shows how internal subjective dispositions reach outward to inhibit change beyond the subject. Thus, increased sensitivity to the shifts of interruptive experiences needs to be fused to a wariness of convention and a willingness to break free of comfort and routine. Here too, Nietzsche makes transition a goal: "There are so many experiments still to make! There are so many futures still to dawn."[18] Repetitive events are energizing because they are unsettling, transmuting tension into creativity.

Nietzsche's analysis of these tendencies reaches it culmination in the Dionysian festival, where he clarifies the experience of the event and connects it to broader cultural forces. Dionysian tragedy and festivals are repetitious and form an intimate link with the renewing forces of spring.[19] The Dionysian embodies a "vivid" and "epic event" that incites a transformation and trading of roles with other cultural elements such as the Apollonian, religion, or science.[20] Nietzsche describes this as a "to-ing and fro-ing" and the "periodic exchange of honorific gifts," serving as a source of periodic rebirth.[21]

The evening disposition does not just provide comfort but ward off the loss of control that the event produces. Nietzsche describes the force of the Dionysian event as a combination of horror and ecstasy that arises when the subject as a consolidated and stable form breaks down.

> the enormous *horror* which seizes people when they suddenly become confused and lose faith in the cognitive forms of the phenomenal world because the principle of sufficient reason, in one or other of its modes, appears to sustain an exception. If we add to this horror the blissful ecstasy which arises from the innermost ground of man, indeed of nature itself, whenever this breakdown . . . occurs, we catch a glimpse of the essence of the *Dionysiac*, which is best conveyed by the analogy of *intoxication*. These Dionysiac stirrings, which as they grow in intensity, cause subjectivity to vanish to the point of complete self-forgetting . . .[22]

The Dionysiac contains an element of horror because it punctures established ways of understanding the world, yet it also contains a reciprocal ecstatic feeling. For Nietzsche, the key is to adopt an ethos that accents the latter element to value to overall transformation. As preconceived understandings rupture, previously obscured elements of the world are experienced more directly or with more attention than before. This imbues experience with feelings of freedom and possibility that are both exciting and frightening. Nietzsche refers to this variously as the "breaking-asunder of the individual," "the playful construction and demolition of the world of individuality," and as an event in which "subjectivity disappears entirely."[23] This destruction brings with it "an energy utterly alien to the placid flow of epic semblance" that "gives birth again and again."[24] Just as spring brings rebirth, so does the Dionysian event. It brings one into the "experience of seeing oneself transformed before one's eyes and acting as if one had really entered another body, another character."[25] Like Kierkegaard, Nietzsche thinks that the event incites existential change. But the transformations that Nietzsche promote are temporary; the rebirths, multiple.

The transformations are temporary because the individual itself is not a natural basic form but a temporary construction. Nietzsche embraces the individual as subject to constant flux, As we shall see, he goes so far as to suggest that the conscious individual subject is a fiction imposed to give coherence to a collection of bodily and affective forces sorted and stabilized by cultural codes and imperatives. The way he uses the event to help us recognize this underlying nature of the individual is critical for climate change responsiveness. Individualism plays an important role in supporting climate change denial as well as coopting those who are trying to respond.[26] Thus, harnessing events to interrupt the culture of individualism and the self-affirmation of particular individuals is critical for the construction of a subject capable of engaging climate change. Such interruptions might also play a key role in the difficult and open-ended project of constituting the responsive "we" of climate change.[27] Because the process of forming a subject is never complete, Nietzsche emphasizes the repetitive aspect of transformations by drawing on seasons, times of day, and festivals.

These transformations, when they are salient enough to organize life, take on the status of myths. Yet as such, they are also in danger of becoming ossified. Just as the evening disposition tries to secure itself through veneration, sometimes a myth supports itself with fact to become a religion.[28] Then, the disposition toward Dionysian events, transfiguration, and the act of creating myths is suppressed in the name of a view that seeks to systematize experience. Particular myths are thereby given historical legitimacy while the process of myth creation is seen as no longer credible. For example, Nietzsche argues that the philosophies of Euripides and Socrates worked to suppress the Dionysiac element in Greek society and that science followed a similar path in his time. Nonetheless, Nietzsche thinks that every systematic edifice will meet its own limits.[29] They also tend to be more fragile and shatter when disruptive events do break through, having sacrificed malleability for cultural legitimacy. While the stable view of a systematic life may be metaphysically comforting, Nietzsche suggests the importance of finding comfort in the tragic interruptions that punctuate life. "You should first learn the art of comfort *in this world*, you should learn to *laugh*, my young friends . . . perhaps then, as men who laugh, you will some day send all attempts at metaphysical solace to Hell."[30] There is no final comfort or security. In a world in flux, comfort, like laughter, is temporary. To insist on more is to look away from the vital changes underway.

The eternal return of affect and the emotional denial of climate change

Nietzsche deals with the repetition of the event more thoroughly through his notion of The Eternal Return. Though it is one of the most widely interpreted concepts in Nietzsche's philosophy, it has seldom been read in terms of the event.[31] A provocative reading is given by Pierre Klossowski in *Nietzsche and the Vicious Circle*, which deepens the themes set out in the previous section. Here I will argue that an active engagement with the affective life of the subject enables interruptive experiences that both de-individualize the subject and rework its connections to established cultural and institutional routines. The effectiveness of this level of engagement can be seen in the way a personal breakdown forged a stronger connection to climate change for meteorologist Eric Holthaus. The importance of it is underscored by Kari Marie Norgaard's study of emotional climate denial.

Klossowski argues that the Return recasts the role of the philosopher, whose task is to channel it.[32] "Does not this event, which the philosopher apprehends . . . first have to be *mimed*, in accordance with the gestural semiotic of the Soothsayers and the Prophets?"[33] The philosopher's intervention is to experiment with different ways of repeating the event to make it tangible, enacting in thought and behavior. It is experimental because it cannot be embodied in a consistent way, as seen in how the event of the Return interrupted Nietzsche's life at the affective level. Reading Nietzsche's diaries and letters, Klossowski suggests that "the thought of the Eternal Return of the Same came to Nietzsche as a [sic] *abrupt awakening* in the midst of a *Stimmung*, a certain tonality of the soul. Initially confused with this *Stimmung*, it gradually emerged as a thought; nonetheless, it preserved the character of a revelation—as *a sudden unveiling*."[34] The actual event cannot be conceptualized. It is a soulful, affective experience, an intensity that dissolves the subject and interrupts conceptualization. Nonetheless, when thought begins to organize and conceptualize the experience, it attempts to preserve the sudden and revelatory character of the event.

The point of the Return is not, however, fidelity to a single revelatory moment, but repeated attempts to reengage and reactivate it. These attempts come up against established moralities.

> We must break with the classic rule of morality, which—on the pretext of realizing a human potential—makes humanity dependent upon habits adopted *once and for all* . . . Behavior can never be limited by its regular repetition, nor can it limit thinking itself. A mode of thought that would restrict behavior, or a mode of behavior that would restrict thought—both comply with an extremely useful automatism: they ensure *security* . . . By contrast, any

thought that allows itself to be called into question, whether by an internal or external event, reveals a certain capacity for starting over."[35]

Two responses to the event and two forms of repetition are distinguished. On one hand, Nietzsche breaks away from a morality that limits thought and behavior by imposing a regular repetition upon the world in the name of improvement and security. This approach suppresses and contains events to reinforce this morality. Alternatively, his approach is one that welcomes internal and external events to see how a morality of regular repetition is actually only a "provisional state."[36] Two different responses correspond to these approaches. The first experiences unease when confronted by events that unsettle established morality. The second uses the event as an opportunity to call morality into question. This response suggests a different model of repetition in which the security of an established morality is abandoned to begin again.

The way that this event manifests is not limited to the experience of the subject but is also connected to the social relations within which the subject exists. Klossowski links that which relates to the Return to the idea of the "singular," while that which relates to language, comprehension, institutionalization, and normalization is linked under the idea of the "gregarious." The event has the power to call gregarious sociality into question through the singular impulsive state in the subject, and so "can bring about a de-actualization of that institution itself."[37] An experience of the intensities that are becoming in the present, or the experiential return to a past impulsive state, can dissolve the social organization of experience, calling into question instilled norms of what ought to be felt, along with the meaning of those feelings.

It is not just that the subject now feels and thinks differently but that it is already behaving differently within the institutionalized context. "*As soon as we act practically . . . we have to follow the prejudices of our sentiments.*"[38] Most sentiment organizes the flux of intensities according to gregarious principles. This inclines behavior toward established social and moral codes. Thus, it is a question of how to rework sentiment, how to take part in the organization of intensities in favor of the singular. "A purely experimental doctrine of selection will be put into practice as a 'political' philosophy . . . [which] can also be regarded as an *invented* simulacrum in accordance with one of Nietzsche's *phantasms*."[39] The new selections are then introduced into gregariousness through the sensibilities and practical activity of the experimenter. Inventing a simulacrum in accordance with a phantasm is a matter of producing an invented constellation of intensities, since the intensities producing the phantasm (the high tonality of soul) in the event resist comprehension.[40] The simulacrum is a manufactured set of intensities that uses the disruption of the event to challenge the institutionalized channeling of experience. The production of simulacra is a "practice" that takes up "a positive notion of the false" which can "generate new conditions of life."[41] Reorganizing the sentiments is thus "an exercise in continually maintaining oneself in a discontinuity with respect to everyday continuity."[42] This distance creates further possibility for transformation.

Through the active production and embodiment of simulacra, the event has its effects, which in turn repeat the event through their own efficacy. "Thought must itself have the same *effectiveness* as what happens *outside of it* and *without it*. This type of thought, in the long run, must therefore *come to pass* as an *event*."[43] Nietzsche does not just seek an idea of Eternal Return that enables one to evaluate life, it must become life. He wants to insert the Eternal Return as an event on the same register as the compulsion to continue producing institutions that "improve" the species but as a force that interrupts it. "[T]he 'Vicious Circle' . . . not only turns the apparently irreversible progression of history into a regressive movement (toward an always undeterminable starting-point), but also maintains the species in an 'initial' state that is entirely dependent on *experimental initiatives* . . . In the course of events, the Eternal Return, as experience, as the thought of thoughts, constitutes the event that abolishes history."[44] The experience of the Eternal Return broke into Nietzsche's life and became an uncertain starting point to which he would continue to return and to which his philosophy urges us to continue to return. It "abolishes history" because returning to the experimental starting point cannot be worked into the gregarious progressive history of the human species.

The efficacy of this repeated interruption occurs on three levels: the subject, the inter-subjective, and the world. At the subjective level, "the agent *unmakes* and remakes itself in accordance with the receptivity of other agents—agents of comprehension."[45] These other agents can be people or institutions, but what is important is that they operate through and impose gregarious meaning upon the subject. Though others are involved at this level, the event manifests itself only in the subject. Nietzsche adopted the image of the mask to enact this effect.[46] With the mask, one wears their own identity or ego as a chance production, no longer directly living that person, but living at the limit between the singular and the gregarious. This is also the limit between the creation and dissolution of the self, the flux of impulses and the gregarious institutions. Rather than experiencing institutionalized subjectivity, one experiences the suggestion of that subjectivity out of the impulses. It is then possible to attend to the way that various institutions solidify and manipulate that subjectivity.

Klossowski looks at letters and diaries of Nietzsche's acquaintances to show that there is an intersubjective dimension to the event, even if it cannot be gregariously articulated. The close friends with whom Nietzsche shared the Eternal Return were often confused by the idea and thought that he was referring to a system of thought from antiquity. Yet there was also an affective dimension which accounts for "the impression of strangeness felt by his friends."[47] "Overbeck emphasizes the state Nietzsche was in when he spoke with him (bedridden, suffering from a migraine), the disturbing tone of his hoarse voice, the spectacular character of the communication."[48] It is as if the affective dimension of the Return was expressed in the intensity of Nietzsche's person when he tried to share it, "and they *felt* the delirium."[49] The combination of an obscure experience that exceeds doctrines with Nietzsche's own affective performance was read by Nietzsche's friends as a symptom of a defective mental state rather than as two sides of a conceptual experience. Klossowski argues that this was an effective inter-subjective enactment of the event. "When Nietzsche invited them to think with him, he was really inviting them to feel, and thus to feel his own prior emotion."[50] In inviting them to feel, he tried to show them how thought comes from the impulses, the same insight he had when he experienced the Eternal Return.

The Eternal Return also has worldly effects. As Klossowski says of Nietzsche: "*He would incarnate the fortuitous case.* At the same time, he would *reproduce* the world, which is merely a combination of random events. Thus he would train himself in the practice of the unforeseeable."[51] As one becomes a fortuitous case, the world becomes a fortuitous case as well, subject to the flux of intensities. Nietzsche enacted, experimented, and projected the contingent world in a way inaccessible and unforeseeable to teleological, deterministic, or purposive worldviews. Experiencing the world as a "*discontinuity of intensities*" frees thought from socially imposed limits and helps to see the possibilities that it contains.[52] Yet living at this limit carries risks. Klossowski suggests that Nietzsche's collapse in Turin produced a "*transfiguration of the world*" so extreme that it was "impossible to live."[53] He diverges from many readers of Nietzsche in seeing his collapse as connected to the Eternal Return as well as his earliest experiences and insights. In Nietzsche's last miming of the Return, he slipped into an unmediated experience of intensity, wherein what coherence there was, was established by institutional authorities. While the former way of recreating the world through the practice of the unforeseeable is productive for responding to the event, the latter shows the efficacy and danger it may hold under extreme conditions. The Eternal Return is a sudden event that manifests subjectively, inter-subjectively, and worldly. Each occurrence is an opportunity to refashion institutionalized language, knowledge, experience, and behavior.

***

On September 27, 2013, after an all-night session, The Intergovernmental Panel on Climate Change released a report detailing the current state of climate science and its predictions for the future. Later that morning, meteorologist Eric Holthaus published a short article on the report entitled: "The world's best scientists agree: On our current path, global warming is irreversible—and getting worse."[54] He found it a straightforward, perhaps even routine task to convey the information contained in that report: that humans cause global warming, that severe impacts are on the horizon, that geoengineering is not an option, and that something must be done immediately. Later in the day, he began "thinking about the report more existentially. Any hope for a healthy planet seemed to be dwindling, a death warrant written in stark, black-and-white data. It came as a shock."[55] So, after switching from a scientific, analytical, and journalistic mode of thinking to an existential one, he fell into shock. The data printed across the page was transfigured into a death warrant in which everyday human action authorized sovereign nature to execute the species.

What then happened in a boarding area in San Francisco International Airport has been described as an "epiphany,"[56] and a "meltdown."[57] Holthaus himself calls it a "hopeless moment."[58] While talking to his wife on the phone, he suddenly found himself weeping. Shortly afterward, he sent the following tweet: "I just broke down in tears in boarding area at SFO while on phone with my wife. I've never cried because of a science report before. #IPCC"[59] Two minutes later, he tweeted: "I realized, just now: This has to be the last flight I ever take. I'm committing right now to stop flying. It's not worth the climate.[60] The following tweets document mixed emotions, consideration of a vasectomy, and a willingness to go extinct.[61]

Efforts to reduce his carbon footprint were not new to Holthaus. He already engaged in green behavior such as recycling, turning off the lights, and using reusable bags. He had also adopted a couple of more substantial commitments: being vegetarian and car-sharing. But he still traveled extensively by plane.[62] The dangerous effects of climate change were also known to him. Holthaus gained a lot of notoriety for his reporting during Hurricane Sandy. His coverage was notable for the links he drew between the storm and climate change.[63] Before his transformation, he thought he was acting responsibly yet at the same time he knew it wasn't enough. Why did someone who knew the dangers of climate change and who had already taken steps to live more responsibly suddenly make such a dramatic commitment?

Klossowski's reading of Nietzsche on the Eternal Return may help to clarify Holthaus's reaction. The way that the IPCC report interrupted Holthaus's life is just one of the many interruptions that climate change entails. His response seems to mimic that interruptive force. In this way, Holthaus's commitment is a break similar to Nietzsche's refusal of established morality. Holthaus broke away from a customary morality that limits thought and behavior in the name of security. He notes three ways that his commitment unsettles secure routines: the discomfort of giving up forms of leisure dependent on long-distance travel; the potential loss of a job that requires one to fly; and the impact on the economy if many people were to dramatically curtail their flying habits or consumption in general.[64] Institutionalized comforts, vocations, comprehension, habit, and norms became de-actualized in Holthaus's intense moment of hopelessness.

Yet this interruptive experience also spurred Holthaus to different actions that embodied the intensities of that emotional state. Sentiment organizes the play of intensities according to established moral codes, prejudicing behavior in favor of regularized sociality. Klossowski suggests that Nietzsche reorganized his own sentiments through producing simulacra as invented constellations of intensities. These simulacra imitate the intense experience of the event to reorganize sentiment. Perhaps Holthaus's commitment can be understood as a simulacrum that reproduces the event. That commitment continues to interrupt established ways of living, compelling him to reorganize his vacations, professional life, and activism.

Because Holthaus's transformation is lived, it also becomes part of the experience of those in contact with him. Holthaus's friends, relatives, and professional associates have to consider reorganizing their lives to adapt to his commitment. He also pushed a broader group of people to confront climate change. This can be seen in the responses to Holthaus's twitter announcement. Some people expressed support or admiration; others felt compelled to make pledges to cut their carbon footprint. Yet others expressed disdain, suggesting that Holthaus was a "beta male" or that he should commit suicide.[65] Other climate writers suggested that his emotional reaction had compromised his professionalism and objectivity, or that he was overreacting to the IPCC report.[66] Klossowski notes a similar effect when Nietzsche tried to tell his friends about the Eternal Return. They failed to see a connection between the idea and the affect. In the same way that Nietzsche invited his friends to think and feel with him, Holthaus invites us to feel the disruption he felt. Yet the most common response was to refuse that invitation and to suppress the disruption by affirming established morality. An isolated individual emotional experience is nonetheless able to effectively extend itself into institutionalized life, embodying the event as a way to disrupt and reorganize those institutions.

One thing that this suggests is that we may be able to better understand our relation to climate change by looking at how Holthaus's declaration makes us feel. Amidst various affirmations and resistances to Holthaus, only some were able to work that connection to climate change into their own lives, making commitments to reduce their carbon footprint and become more politically involved. Whether our own organization of sentiment leads us to feel joy or anger at Holthaus's commitment does not matter if either functions to purge the interruptive discomfort of climate change.

***

It is important to understand not just how emotional interruptions can connect people to climate change, but how gregarious morality produces climate denial at the emotional level.[67] In *Living in Denial*, Kari Marie Norgaard examines emotional life to better understand how climate denial is organized, with particular attention to the case of Norway. Norway is particularly intriguing because education levels, widespread knowledge of the issue, and a culture of humanitarian action and political involvement make it an unlikely case for climate denialism. Yet the possession of large petroleum reserves and a relatively high per capita carbon footprint position carbon at the center of society, opening numerous routes for political action and nonaction. This creates a conflict between what is known and what is done that serves as a fruitful site for exploring how emotional life can undermine action on climate change. Norgaard goes beyond individual emotional life to look how emotions are guided by social norms, resulting in socially organized denial.

Similar to Nietzsche, Norgaard argues for the centrality of emotions in directing action. For her, denial and apathy do not indicate a lack of care but rather should be understood "synonymously with *nonaction* and *nonresponse*."[68] Apathy is not a lack but is actively produced and thus always itself a kind of action resulting from a particular social and political context. Indeed, apathy may even be a "mask" that covers care and other emotional investments in a situation that is "too intense."[69] On one hand, Norgaard's conversations with people show that it is fairly common to express "feelings of deep concern and caring and a significant degree of ambivalence about the state of the world."[70] But sometimes other emotions come forward that are more difficult for people to confront or discuss, such as "fears for the future, feelings of helplessness, and feelings of guilt."[71] Norgaard draws two conclusions from this. First, the problem with acting on climate change is not an information deficit, since people are clearly aware of the issue and affected by it. Second, the denial that prevents action occurs at the emotional level. The question is what causes the suppression of intense emotions and blocks action guided by care?

Norgaard analyzes the problem of denial rooted in emotions through what Anthony Giddens calls "ontological security," which "refers to the confidence that most human beings have in the continuity of their self-identity and the constancy of their surrounding social and material environments of action."[72] Norgaard addresses an issue similar to that of Nietzsche by attending to the emotions that are suppressed in the name of a normalized framework of secure identity and action. Norgaard further emphasizes, however, the social sources of these norms and the way they act as a kind of political control within the individual. Norgaard analyzes a variety of "tools of order" and "tools of innocence." The former "affirm a sense of how things are in the world, their ontological stability" and include techniques like joking, focusing on facts, referring to tradition, a sense of place, or connections to nature.[73] The latter "create distance from responsibility and assert rightness and goodness of actions" and include references to a mythic image of Norway, the relatively small influence of Norway geopolitically, comparisons to America, and past suffering.[74] In each case, the effect of such tools is to manage emotion in a way that prevents in-depth discussions about the role of Norwegians and Norway in climate change as well as actions to respond to this environmental and humanitarian crisis.

The socially produced set of emotional states which are acceptable to experience such as being cool, tough, and in control override the initial feels of concern and underlying feelings of guilt and fear. It is not just emotion that is silenced here but knowledge as well. What Norgaard uncovers is individuals actively "*resisting* available information and doing so both *because of* and *through* social norms and interactions."[75] The political implications extend from the social production of denial at the individual level of feeling, thought, and action, to the reinforcement and normalization of national policies that help drive climate change. In addition, the denial itself perpetuates an uneven system of environmental privilege in which some can afford to be in denial about environmental degradation while others will suffer as a result.[76] For these reasons, Norgaard sees the management of emotions as the primary factor solidifying inaction on climate change.

The emotional register must be engaged not only for the obstacle it forms but also because of the opportunity it presents, since the powers of emotions flow in the other direction as well. "Emotions may unite people in common cause, inform their interpretations of the world, and thus catalyze both the sociological imagination and political power."[77] It may be that Nietzsche's theory of the event is helpful for this task. His theory of the Eternal Return helps give voice to suppressed affects that can burst forth and then be mimed in an attempt to replicate that interruptive effect against the secure norms in order to change the direction of politics and society, as in the case of Eric Holthaus. The next section takes up a second of Nietzsche's strategies for altering the way drives below the level of consciousness influence action.

Drive and experiment

Nietzsche sees the subject as situated between affects that it is not entirely conscious of and social norms that play a strong role in determining which affects get expressed in the individual and how they do so. How, then, can the subject be consciously modified? So far, I have looked at the event as an extreme state that dissolves and transforms the subject. While this is an important component of Nietzsche's thought, he also seeks to consciously shape the drives in a less intense, open-ended fashion. To do so, he attends to the relationship between the drives and everyday experience, habits, customs, and dispositions. Here, the event takes the form of an experimentally adopted "brief habit" which is valorized only to be given up for a new one to see what effect this produces on the organization of the drives. These experimental transformations are thus a creative way to shape a self that is not subject to rational control.

First, it is necessary to briefly sketch Nietzsche's theory of the drives. He clarifies the overlooked role of the drives in a section of *Daybreak* entitled "Experience and invention." "However far a man may go in self-knowledge, nothing however can be more incomplete than his image of the totality of *drives* which constitute his being. He can scarcely name even the cruder ones: their number and strength, their ebb and flood, their play and counterplay among one another and above all the laws of their *nutriment* remain wholly unknown to him."[78] People rarely think about the structure of their experience, and even when they do, they lack the resources to account for it, in part because they are inexperienced in this art and in part because only some of the drives reach up to conscious experience. The way in which drives get expressed is not a matter of the strongest or most hungry drive triumphing over the others. Rather, it is "a work of chance: our daily experiences throw some prey in the way of now this, now that drive, and the drive seizes it eagerly; but the coming and going of these events as a whole stands in no rational relationship to the nutritional requirements of the totality of the drives."[79] Thus, the satisfaction of the drive can mean many things: it "desires gratification—or exercise of its strength, or discharge of its strength, or the saturation of an emptiness—these are all metaphors."[80] The problem of the drives is not epistemological, but ontological: there is no rational or systematic way in which the interplay of drives occurs. Moreover, the satisfaction of one drive through a particular event does not produce an overall state of satisfaction but a different configuration of feeling and intensity. This is why Nietzsche sees it as a matter of chance. Nietzsche thereby calls us to go beyond what we normally take to be experience and attend to the drives that shape it.

Nietzsche also extends the influence of the drives upward into the higher levels of cognition. "Our moral judgments and evaluations too are only images and fantasies based on a physiological process unknown to us, a kind of acquired language for designating certain nervous stimuli . . . all our so-called consciousness is a more or less fantastic commentary on an unknown, perhaps unknowable, but felt text."[81] The drives are the foundation for what we often take to be the height of conscious determination: our values. The attraction or revulsion we experience when we see something we judge to be base or noble is the due to the drives. Nietzsche goes so far as to attribute his own spontaneous selfless acts to the drives and even more than he would not have performed those actions if he had known what problems he would encounter in advance.[82] Thus, the drives are not deterministic but set the subject in the world in a partial and differential participation way, with consciousness playing its own incomplete part. This is why Nietzsche suggests that "to experience is to invent."[83] Primary agency is accorded neither to the conscious subject nor the drives that partially structure it, nor the world they inhabit. Experience is thus not determined on any particular level but a creative emergent invention based on the chance composition of the moment.

Drives are not simply blind impulses in the individual. They have roots in culture, social organization, politics, and history. To understand "how differently the human drives have grown and still could grow" would require an examination of labor practices, moral climate, nutrition, organization of time, law, festivals, and much more.[84] Nietzsche concludes that the last step in the project of researching the drives would be to see "whether science is able to *furnish* goals of action after having proved that it can take such goals away and annihilate them; and then an experimenting would be in order."[85] Nietzsche undermines the assumed legitimacy of dominant moral objectives by showing them to be cultural, historical, and affective: in other words, invented and temporary. But he also turns this around to ask what future modes of being are possible if we turn this into a project of experimentation. Experimentation works on the unknown but indirectly observable composition of the drives. But rather than being a reason to pass over the drives, their shifting relation to consciousness makes them fertile ground for reshaping human experience.

Nietzsche does not want this to be understood as a planned set of experiments for human betterment or teleological progress. His concern is how experience is constituted now and how it might be constituted next. This assumes that humans have lived in many other ways before and that they will live in yet other ways in the future. Nietzsche calls our world a "moral interregnum." "So it is that, according to our taste and talent, we live an existence which is either a *prelude* or a *postlude*, and the best we can do in this *interregnum* is to be as far as possible our own *reges* and found little *experimental states*. We are experiments: let us also want to be them!"[86] We cannot escape the fact that our lives are not building blocks in a progressive teleological history but rather between-periods. In critical moments like the Anthropocene, there is every temptation to fall back onto some kind of overarching narrative, whether it be global collapse because of hubris and greed or a bright recovery thanks to reason and green technology. But if we come to appreciate not being subject to deterministic historical laws and moralities, we can explore the possibilities that exist for shaping open-ended iterations of them. Then, "life could be an experiment for the knowledge-seeker . . . *Life as a means to knowledge*."[87] Experimentation becomes immanent to experience and itself.[88]

In this situation, each experiment reshapes experience in such a way that it produces the possibility for further experimentation. In this way, Nietzsche emphasizes the transitional and transformational element of moving between different states of being that was so important for him with regard to seasons, festivals, and the Eternal Return. And perhaps nothing is more important today that developing a capacity for transformation as the only thing that emerges through the growing crisis is how set in our ways we have become.

Nietzsche's most compelling formulation of this experimental disposition is his notion of brief habits. "I love brief habits and consider them invaluable means for getting to know *many* things and states down to the bottom of their sweetnesses and bitternesses . . . I always believe *this* will give me lasting satisfaction—even brief habits have this faith of passion, this faith in eternity."[89] In the form of a habit, Nietzsche makes the solidity and intensity of a dominant moral system into something transitory. An almost Kierkegaardian language is used to describe the investment in and fulfillment from an undertaking that one treats as eternal even though one knows it will soon be left behind. Each habit brings with it a different kind of existence for the experimenter, and each time a habit is given up and another taken on, experience and drives are reconfigured. In this way the experimenter comes to know the advantages and disadvantages not just of "habits," but also of modes of existence that have enough consistency to nourish the self.

It is not just a matter of the particular habits but also the experimental ethos itself. Nietzsche continues: And one day [the habit's] time is up; the good thing parts from me, not as something that now disgusts me but peacefully and sated with me, as I with it, and as if we ought to be grateful to each other and so shake hands to say farewell.[90] Despite the passion with which Nietzsche commits to each habit in order to explore it, he does not resent either its failure to be an answer or its passing: he is content with the knowledge and experience gained. Thus he refigures the interruptive character of the event. The break between the ways of being that structure life does not arrive as a collapse or downfall but as a grateful parting. This is an ethos of the event that transfigures the negative affect of collapse into a positive valuation of the next experiment" This is the eventful aspect of experimentation: the way that a person undergoes a transformation of their constitution. To continually do so requires an ethos that commits to each way of life, tries many of them, and leaves each cheerfully behind. The experimental disposition also helps Nietzsche guard against enduring habits or permanent modes of life, which he sees as tyrannical.

Through these experiments, one can indirectly influence the structure of the drives or as Nietzsche calls it, "'give style' to one's character." "Here a great mass of second nature has been added; there a piece of first nature removed—both times through long practice and daily work at it."[91] The "first" nature is not more fundamental or essential than the second; each part of our constitution can be reshaped, augmented, removed, or refined. Each experiment will result in a change in the configuration of the drives. This change will then filter up into consciousness and experience. One can't say ahead of time what the outcome will be. We may even be upset with the kind of person we become and see the experiment we committed to as an error. "But maybe that error was as necessary for you then, when you were still another person—you are always another person . . . there are living, active forces within us shedding skin."[92] Reason will never arrive at a position from which it can compare the various ways of being to form a final judgment on them. The way that the drives and reason run together to constitute the self means that any new realization for reason is always part of a change in the drives. What might emerge, however, is a greater sensitivity to the movements in the self, the ability to commit to unknown changes, and a joy that infuses this process.

Experimentation also faces resistances similar to the dangers of the evening disposition and ossified myths. In particular, Nietzsche is concerned that the disposition against becoming may frustrate experimentation upon the drives.[93] A stable morality of causal determinism, for example, might be used to suppress experimentation and thus the instability of becoming.[94] It is not just a matter of overcoming the impulse to subordinate experience to reason, the problem is also to negotiate predispositions that mobilize fear, anxiety, and comfort in such a way that experimentation becomes an uncomfortable process. If humans are part of a world of becoming, then neither they nor any other cause can be isolated as a singular, durable, and self-sufficient center of organization. There is no external measure by which to judge what occurs but rather experience judges itself. This view favors provisional judgments and continued experimentation. As Nietzsche puts it, "when we speak of values we do so under the inspiration and from the perspective of life: life itself evaluates through us *when* we establish values."[95] This is not a completely worked out framework to follow. It is a condition to begin experimenting.

In contemporary society, economic growth serves as a stable morality that suppresses becoming and experimentation. Clive Hamilton argues that growth does not follow rational principles of economic evaluation but has become a "religious urge," a "fetish," and an "*unreasoning* obsession" with how to increase well-being, even though further growth does not make people in wealthy nations happier.[96] Thus is it not growth as such, but the growth fetish that is "the principal obstacle to effective global warming policies."[97] This problem is deepened by the way consumption and materialism have become a critical component of individual identity formation. Marketing firms seek to "play on complex sets of feelings" to get people to buy certain products by playing on fear and dissatisfaction while promising fulfillment gratification, and authenticity.[98] This has driven an extreme form of overconsumption that persists even though people know it is wasteful and feel anxious, guilty, or depressed about it.[99] Green consumerism has allowed some individuals to allay those feelings by offering a more moral identity, though it does not change the underlying dynamic of overconsumption.[100]

Thus growth fetishism combined with the construction of consumer identities presents a powerful moral imperative that has taken root in the unconscious drives of individuals with devastating effects for the climate. Nietzsche's engagement with drives deepens our understanding of how Hamilton tries to sort through this problem. When Hamilton argues that efforts to minimize the impacts of climate change amount to a "war against our own sense of who we are," this means that that war needs to be engaged at the complex and contradictory level of the drives.[101] If Hamilton is right in suggesting that this means we have "to experience a sort of death," then Nietzsche helps us see how the task of constantly remaking oneself without any demand for permanent identity is a positive one.[102] Hamilton seems pleased with the exercises in prudence, moderation, and delayed gratification that some people were forced to take up in the wake of the 2008 financial crisis, but despairs over how quickly that energy was directed back into growth activities as the economy recovered. Thus, Hamilton suggests a set of deeper cultural-economic changes that include lower working hours, more hobbies and leisure, and less income. Hamilton argues that this would require a shift in identity, though does not propose a way to do that.[103] Nietzsche might try different experiments along these lines to begin the necessary shift in identity through reconfiguring the interplay of the drives. This experimental disposition would also be more apt to view this as only a starting point that can't solve the climate problem but open up paths through which to further positively engage it.

Becoming emigrants: the death of god and the death of the human in the Anthropocene

So far, I have looked at the subjective and social registers of the event in Nietzsche's thought to suggest how they might be brought together in composing responsive subjects to climate change. One more level must still be added. Nietzsche theorizes great events to think beyond the life of the individual, particular societies, and even the species. Great events shape epochs, ecologies, and cultural modes of existing but occur on a timescale that can be difficult to perceive and understand, in part because they are beyond the control of an individual, group, nation, or species. Understanding such events and how they shape life requires adopting a different temporal orientation. This can be seen in Nietzsche's approach to two events: the human species and the death of God. Such large-scale events help us come to terms with our position in the Anthropocene with its planetary shifts, extinction events, and unknown trajectory.

Nietzsche assigns the task of "learning solitude" to those who want to understand great events. It begins by distancing oneself from the normal frame of reference for human action. "O you poor devils in the great cities of world politics, you gifted young men tormented by ambition who consider it your duty to pass some comment on everything that happens . . . However much they may desire to do great work, the profound speechlessness of pregnancy never comes to them! The event of the day drives them before it like chaff, while they think they are driving the event."[104] Superficial responsiveness is an obstacle to engaging great events. It reduces the event to an extension of the ability of the individual or group, rather than recognizing the ways that the event permeates life at a deeper level. In seeking to have an effect, those bent on exercising their wills do not guide the event but are symptoms of it. Al Gore, the military, and green energy companies all try and act on climate change, but their actions are reactions that fail to take into account how this event already courses through them as well as the broader Earth system. Have they come to terms with planetary feedback in the Earth system that operates beyond human intentionality?[105] Have they really tried to understand how carbon energy has captured and guided human development?[106] To be able to influence great events, politics may have to become silent, learning how to listen and exist beyond an individual's or society's time and place.

Turning away from noisy and human-cluttered visions of the event, these questions may be engaged through Nietzsche's suggestion to learn silence as a way of moving outside the human frame of reference. In a section entitled "In the great silence," Nietzsche describes the experience of being away from the city and by the sea at twilight. Bit by bit, one realizes how silent everything is, how it does not speak: first the sea, then the sky, then the cliffs. Yet one feels malice in this silence.

> Ah, it is growing yet more still, my heart swells again: it is startled by a new truth, *it too cannot speak*, it too mocks when the mouth calls something into this beauty, it too enjoys its sweet silent malice. I begin to hate speech to hate even thinking; for do I not hear behind every word the laughter of error, of imagination, of the spirit of delusion? Must I not mock at my pity? Mock at my mockery?—O sea, O evening! You are evil instructors! You teach man to cease to be man! Shall he surrender to you? Shall he become as you now are, pale, glittering, mute, tremendous, reposing above himself?[107]

Humans are invested in speaking, thinking, and producing truths and yet this silence shows they do not matter to nature. The division between man and nature is, however, not so easy. In the silence, Nietzsche notices that his own heart becomes silent. The nature in him does not so much speak up as make its silence felt. Nietzsche discovers the larger natural forces operative in and coursing through his own existence, which pulls his mind away from consciousness and human intention. Thus, the sea can "teach man to cease to be man." In a way, even our hearts are not ours. It can carry us to the limit of our human-centered views, enabling us to examine our own conditions of possibility on larger timescales, even that of the species. Learning silence clarifies the human condition, opening different approaches to the world in which it is set.

Nietzsche begins his essay "On Truth and Falsity in their Ultramoral Sense" with a starker rendering of the event of the human.

> In some remote corner of the universe, effused into innumerable solar-systems, there was once a star upon which clever animals invented cognition. It was the haughtiest, most mendacious moment in the history of this world, but yet only a moment. After Nature had taken breath awhile the star congealed and the clever animals had to die.—Someone might write a fable after this style, and yet he would not have illustrated sufficiently, how wretched, shadow-like, transitory, purposeless and fanciful the human intellect appears in Nature.[108]

On a cosmic timescale, the birth and death of humanity as a great event takes only a few of Nature's breaths.[109] Even the lifespan of the planet is short and will come, whether humans dramatically change its climate or not. The species endeavor with the intellect at its center is one cosmic event among many. So far, Nietzsche thinks humans have adopted two responses to being situated in nature in this way. The first is the attempt to establish stability and truth through language, concepts, and science so that a dangerous and unstable nature can be met with "foresight, prudence, [and] regularity."[110] The problem with this view is that it refuses to deal with human existence as an uncertain and contingent event, instead explaining and even trying to control nature through a human ordering of it. The false opposition that this establishes between humans and nature will collapse when the conditions supporting human existence start to tremble, as the increasing literature on extinction events suggests.

Nietzsche prefers a more artistic response that "constantly shows its passionate longing for shaping the existing world of waking man as motley, irregular, inconsequentially incoherent, attractive, and eternally new as the world of dreams is."[111] Rather than looking for a reason that justifies existence and secures the human place within it, this way of life embraces the arbitrariness of the world. Instead of building an edifice of laws and knowledge, humans create according to their sense, whim, and intellect such that "everything is possible; and all nature swarms around man as if she were nothing but the masquerade of the gods."[112] Rejecting views that see nature as an object to be studied or an external threat to human life, Nietzsche argues that humans should joyfully take up their contingent connections, limitations, and possibilities within it. Climate change is almost exclusively treated as a series of limits that restrict human freedom and flourishing. But it really only restricts a particular human project at a particular time and only appears as a limitation to those who are unable to see the small event of their own civilization in relation to other human and natural events. Many human cultures and civilizations have existed in fulfilling and beautiful ways that did not provoke this problem. There is no reason to imagine that decarbonized life would not be another attractive creation. From this vantage, even the threat of human extinction is transfigured into a conflicting confluence of partially defined human projects amidst larger natural forces that take them over like gods directing a tragic fate.

Dipesh Chakrabarty pushes in a similar direction when arguing for developing "epochal consciousness," a term he takes from Karl Jaspers.[113] For him, views of climate change like that of the Capitalocene that focus only on differentiation within social systems miss a critical part of what his happening. "But planetary climate change and the Anthropocene are also events driven by nonhuman, nonliving vectors that work on multiple scales, some of which work on geological scales while some have an influence within the time horizon of one or two human generations."[114] At the most basic level, even an analysis of unjust social institutions will miss the coming injustice if it cannot also step back from the social to see the deep time climatic history of Earth and the drivers of those changes. But perhaps more problematic is that "justice arguments are not very good at thinking limits."[115] There is no way to give developing nations the same carbon allowance that developed nations have taken without provoking extreme climate change. The planet may not be amenable to human conceptions of justice. That doesn't even begin to deal with the history of the human species in relation to other forms of life and it's intensification now during the sixth great extinction. For this reason, Chakrabarty argues that epochal consciousness will have to add a zoecentric perspective to its other analysis of climate change.

Even beyond this, we need to develop "recognition of the otherness of the planet itself."[116] This means that the Earth is not just a dwelling place for humans but that it has its own modes of operation and projects that it has set in motion that may not take human existence into account or may even be hostile to it. This is not just a matter of scientific truth of philosophical honesty, but a political question. Epochal consciousness draws our attention to the planet as a cofactor in climatic shifts that affect the conditions of habitability for life.

> Given this phenomenological aspect to epochal consciousness, our affective responses to climate change—from denial to moods of heroism—all seem understandable and will no doubt continue to influence the politics of global warming. Motivating globally coordinated human action on global warming necessarily entails the difficult if not impossible task of making available to human experience a cascade of events that unfold on multiple scales, many of them inhuman.[117]

Without directly attending to it, Chakrabarty links the three different registers of climate change that have been discussed through Nietzsche's thought in this chapter: the affects inside the subject, the politics of social relations, and the planetary. None are sufficient on their own and they are not reducible to one key register. Before climate change, Nietzsche already urged us to think about our fragile and contingent place on the planet. Chakrabarty pushes that forward in connecting our political responses to climate change to the affective response we have to such a position. In this way, a joyful ethos is needed for engaging the trouble of such great events in order to mobilize a more effective and positive politics across a divided humanity.

Nietzsche develops this positive ethos more thoroughly in relation to great events that occur within the human venture. The one that concerns Nietzsche the most is the death of God. He lived in the midst of this occurring-but-not-yet-heard event. "This tremendous event is still on its way, wandering; it has not yet reached the ears of men."[118] Even in the middle of them, such events are hard to perceive, understand, and react to. Someone who recognizes it and takes it seriously may be seen as a "madman." Yet even they are lost within it "What were we doing when we unchained this earth from its sun? Where is it moving to now? Where are we moving to?"[119] To respond to such events, Nietzsche suggests developing a positive ethos that transforms elements of uncertainty and directionlessness into an opportunity to think and change. Then, the immediate consequences of the event become "At hearing the news that 'the old god is dead', we philosophers and 'free spirits' feel illuminated by a new dawn; our heart overflows with gratitude, amazement, forebodings, expectation—finally the horizon seems clear again, even if not bright."[120] Nietzsche tries to generate a positive affective milieu that excites us for something new. Rather than resenting change, it should be taken up with curiosity and energy. Faced with upheaval and an uncertain future, fear and foreboding remain, but they do not paralyze action.

It is not just a question of developing a positive feeling for responsiveness, but the ethos must also affirm a world that contains such uncertain developments. Great events put us into a state of questioning and can push us to critically confront the values we have lived and accorded to things. Here is the audit that Nietzsche gives during the death of God: "... the way of the world is not at all divine—even by human standards it is not rational, merciful, or just. We know it: the world we live in is ungodly, immoral, 'inhuman' ... We take care not to claim that the world is worth *less*; indeed, it would seem laughable to us today if man were to aim at inventing values that were supposed to *surpass* the value of the real world."[121] The problem is not the hardships the world places upon us or of the existential comfort that it denies us, but the values we have produced that distort existence and our expectations of it too much. For Nietzsche, such myths are unhelpful: he prefers those that beautify the sublime and tragic dimensions of becoming in this world rather than those that "beautify" worldly existence through a static image of something fundamentally "better."

The acceptance and perhaps even love of a world capable of climate change is a critical component of responding to it. As Norgaard points out, "the fact that nobody wants information about climate change to be true ... fit[s] perfectly with the agenda of those who generate skepticism."[122] The underlying desire for a different world makes it easy for who do not buy into the climate denial narrative to nonetheless pursue a similar set of actions when it comes to transforming the organization of society. William E. Connolly sharpens this problem into "passive nihilism," or the "formal acceptance of the fact of rapid climate change accompanied by a residual, nagging sense that the world ought not to be organized so that capitalism is a destructive geologic force."[123] Passive nihilism infuses the struggles of everyday life making stronger action on climate change more difficult. Nietzsche helps us work through this problematic disposition first by refuting it in stark terms that we might initially find difficult to accept; second, by posing questions and lines of thought that unsettle our security in that disposition; and third, by giving positive value to an "emigrant" approach that gives up the comfort of home to explore the new. In this way we may be able to love not the phenomenon of climate change, but the world that includes it as an event tied to the human.

Caught in the "broken time of transition," those who have developed a positive ethos of the event find the comforting ideals of stability unappealing because they expect things to continue to change.[124] From this position they can participate in and shape the event. "The ice that still supports people today has already grown very thin; the wind that brings a thaw is blowing; we ourselves, we homeless ones, are something that breaks up the ice and other all too thin 'realities.'"[125] The freer disposition of those who accept the homelessness of the event propel it along, taking up the current of the event as their own. This is particularly critical today, when, as Latour argues, "the migratory crisis has been generalized."[126] His point is not to flatten the real differences of social relations including those who have already lost their territories of existence to climate change, but to point out that the land that people have lived on and taken for granted is itself undergoing transformation: drying out, being submerged, ecologically shifting. While Latour pursues a political solution of finding security beyond national identity and borders, becoming emigrants in the Nietzschean sense may be a critical addition to this project given that he too seeks to draw on affects, social relations, and planetary transformation. He draws not on Nietzsche, but Carl Schmitt to do so, whose approach to the event will be the subject of the next chapter.

Notes

1 While the term "Anthropocene" does not have a singular definition and historical dating attached to it, its main advocates share a number of assumptions that unite them. See Bonneuil and Fressoz, *The Shock of the Anthropocene*, chapters 1–4.
2 Bonneuil and Fressoz, *The Shock of the Anthropocene*, 65–72.
3 Ibid., 72–9.
4 Ibid., 79–85.
5 Ibid., 86–94.
6 Ibid., 93.
7 Latour, *Facing Gaia*.
8 Chakrabarty, "The Human Condition in the Anthropocene."
9 Connolly, *Facing the Planetary*.
10 Moore, *Capitalism in the Web of Life*, 172.
11 Haraway, *Staying with the Trouble*, 51–5.
12 Guattari, *The Three Ecologies*, 17–18.
13 For an important discussion of the problems of using Nietzsche politically, see Waite, *Nietzsche's Corps/e*.
14 Nietzsche, *The Gay Science*, 3.
15 Ibid., 6.
16 Nietzsche, *Daybreak*, 168.
17 Ibid., 214–16.
18 Ibid., 109–10.
19 See, for example, Nietzsche, *Birth of Tragedy*, 19–21; Nietzsche, "The Dionysiac Worldview," 119–24.
20 Nietzsche, *Birth of Tragedy*, 83, 112.
21 Ibid., 28, 20.
22 Ibid., 17.
23 Ibid., 44, 114; Nietzsche, "The Dionysiac Worldview," 120.
24 Nietzsche, *Birth of Tragedy*, 34.
25 Ibid., 43.
26 Hamilton, *Requiem for a Species*; Maniates, "Individualization"; McCright and Dunlap, "Cool Dudes."
27 Chakrabarty, "The Human Condition in the Anthropocene"; Miéville, "The Limits of Utopia."
28 Nietzsche, *Birth of Tragedy*, 53–4.
29 Ibid., 73–5.
30 Ibid., 12.
31 For an exception to this, see Pearson, "Living the Eternal Return as the Event." For a reading complementary to the one developed below but not focused on the event, see Bataille, *On Nietzsche*.
32 Klossowski, *Nietzsche and the Vicious Circle*, 3–4.
33 Ibid.

34 Ibid., 56.
35 Ibid., 4.
36 Ibid.
37 Ibid., 80.
38 Ibid., 121.
39 Ibid., 125.
40 Smith, translator's preface to *Nietzsche and the Vicious Circle*, x–xi.
41 Klossowski, *The Vicious Circle*, 132–3.
42 Ibid., 65.
43 Ibid., 168–9.
44 Ibid., 170–1.
45 Ibid., 218.
46 Ibid., 224–5.
47 Ibid., 213.
48 Ibid.
49 Ibid., 217.
50 Ibid., 218.
51 Ibid., 220.
52 Ibid., 252.
53 Ibid., 238–9.
54 Holthaus, "The World's Best Scientists Agree."
55 Holthaus, "Why I'm Never Flying Again."
56 Bowman, "What Triggers a Climate Epiphany?"
57 Fox News Insider, "Gutfeld Reacts to Meteorologist."
58 Goldenberg, "IPCC Report Makes US Meteorologist Cry—and Give Up Flying."
59 Holthaus, "Why I'm Never Flying Again."
60 Ibid.
61 Olson, "A Weather Man Breaks Down."
62 Holthaus, "Why I'm Never Flying Again."
63 Berger, "Who Is Eric Holthaus?"
64 Holthaus, "Why I'm Never Flying Again"; Holthaus, "I Spent 28 Hours on a Bus."
65 Holthaus, "I Spent 28 Hours on a Bus."
66 Samenow, "Meteorologist Eric Holthaus' Vow."
67 Dale Jamieson, whose earlier philosophical work on climate change was critically evaluated in the first chapter, has since begun to think more about the relationship between emotions and climate change in *Love in the Anthropocene* with Bonnie Nadzam. This is a step forward, though it still neglects the political dimensions of emotional life.
68 Norgaard, *Living in Denial*, 60.
69 Ibid., 58–62.
70 Ibid., 80.
71 Ibid.
72 Norgaard quoting Giddens, 81–2.
73 Norgaard, *Living in Denial*, 171, 215.
74 Ibid.
75 Ibid., 134.
76 Ibid., 208–20.
77 Ibid., 213.
78 Nietzsche, *Daybreak*, 74.
79 Ibid.
80 Ibid.
81 Ibid., 75–6.
82 Ibid.
83 Ibid.

84   Nietzsche, *Gay Science*, 34.
85   Ibid., 35.
86   Ibid., 190–1.
87   Ibid., 181.
88   See Connolly, "Experience and Experiment"; Connolly, *The Fragility of Things*, postlude.
89   Nietzsche, *Gay Science*, 167.
90   Ibid., 167–8.
91   Ibid., 163–4.
92   Ibid., 174–5.
93   Nietzsche, *Twilight of the Idols*, 45.
94   Ibid., 62.
95   Ibid., 55.
96   Hamilton, *Requiem for a Species*, 63–5.
97   Ibid., 22.
98   Ibid., 66–71.
99   Ibid., 72.
100 Ibid., 77–81.
101 Ibid., 75.
102 Ibid., 74.
103 Ibid., 84–8.
104 Nietzsche, *Daybreak*, 107. Translation modified.
105 For an orientation point on this question, see Pearce, *With Speed and Violence*.
106 For an attempt in this direction, see Vollmann, *Carbon Ideologies*.
107 Nietzsche, *Daybreak*, 181.
108 Nietzsche, "On Truth and Falsity in Their Ultramoral Sense," 173.
109 Varieties of apocalypticism have proliferated around the issue of climate change. The most thorough discussion of them is Danowski and Viveiros De Castro, *The Ends of the World*.
110 Nietzsche, "Truth and Falsity," 190.
111 Ibid., 188.
112 Ibid., 189.
113 Chakrabarty, "The Human Condition in the Anthropocene."
114 Ibid., 159–60.
115 Ibid., 172.
116 Ibid., 183.
117 Ibid.
118 Nietzsche, *Gay Science*, 120.
119 Ibid.
120 Ibid., 199.
121 Ibid., 204.
122 Norgaard, *Living in Denial*, 181.
123 Connolly, *Facing the Planetary*, 9.
124 Nietzsche, *Gay Science*, 241.
125 Ibid.
126 Latour, *Down To Earth*, 6.

Bibliography

Bataille, Georges. *On Nietzsche*. Translated by Bruce Boone. London: Continuum, 1992.
Berger, Eric. "Who Is Eric Holthaus, and Why Did He Give Up Flying Today?". *SciGuy* (blog), *The Houston Chronicle*, September 27, 2013. http://blog.chron.com/sciguy/2013/09/who-is-eric-holthaus-and-why-did-he-give-up-flying-today/.

Bonneuil, Christophe, and Jean-Baptiste Fressoz. *The Shock of the Anthropocene: The Earth, History and Us*. Translated by David Fernbach. New York: Verso, 2017.

Bowman, Tom. "What Triggers a Climate Epiphany? With Eric Holthaus." In *Climate Report*, October 26, 2013. tombowman.com/posts/what-triggers-a-climate-epiphany/.

Chakrabarty, Dipesh. "The Human Condition in the Anthropocene." In *The Tanner Lectures in Human Values*. New Haven, February 18–19, 2015. https://tannerlectures.utah.edu/Chakrabarty%20manuscript.pdf.

Connolly, William E. "Experience and Experiment." *Daedalus* 135, no. 3 (2006): 67–75.

Connolly, William E. *Facing the Planetary: Entangled Humanism and the Politics of Swarming*. Durham: Duke University Press, 2017.

Connolly, William E. *The Fragility of Things: Self-Organizing Processes, Neoliberal Fantasies, and Democratic Activism*. Durham: Duke University Press, 2013.

Danowski, Déborah, and Eduardo Viveiros De Castro. *The Ends of the World*. Translated by Rodrigo Nunes. Malden, MA: Polity Press, 2017.

Fox News Insider. "Gutfeld Reacts to Meteorologist Quitting Air Travel Over Climate Report." *Fox News Insider* (blog), September 30, 2013. http://foxnewsinsider.com/2013/09/30/meteorologist-eric-holthaus-quits-air-travel-over-un-climate-report.

Goldenberg, Suzanne. "IPCC Report Makes US Meteorologist Cry—and Give Up Flying." *The Guardian*, October 3, 2013. www.theguardian.com/environment/2013/oct/03/ipcc-climate-report-eric-holthaus.

Guattari, Félix. *The Three Ecologies*. Translated by Ian Pindar and Paul Sutton. New York: Bloomsbury, 2014.

Hamilton, Clive. *Requiem for a Species*. New York: Earthscan, 2010.

Haraway, Donna. *Staying with the Trouble: Making Kin in the Chthulucene*. Durham: Duke University Press, 2016.

Holthaus, Eric. "Why I'm Never Flying Again." *Quartz*, October 1, 2013. http://qz.com/129477/why-im-never-flying-again/.

Holthaus, Eric. "The World's Best Scientists Agree: On Our Current Path, Global Warming Is Irreversible—and Getting Worse." *Quartz*, September 27, 2013. http://qz.com/129122/the-worlds-best-scientists-agree-on-our-current-path-global-warming-is-irreversible-and-getting-worse/.

Jamieson, Dale, and Bonnie Nadzam. *Love in the Anthropocene*. New York: OR Books, 2015.

Klossowski, Pierre. *Nietzsche and the Vicious Circle*. Translated by Daniel W. Smith. Chicago: University of Chicago Press, 1997.

Latour, Bruno. *Down to Earth: Politics in the New Climatic Regime*. Translated by Catherine Porter. Medford, MA: Polity Press, 2018.

Latour, Bruno. *Facing Gaia: Eight Lectures on the New Climatic Regime*. Translated by Catherine Porter. Medford, MA: Polity Press, 2017.

Maniates, Michael F. "Individualization: Plant a Tree, buy a Bike, Save the World?" *Global Environmental Politics* 1, no. 3 (2001): 31–52.

McCright, Aaron M., and Riley E. Dunlap. "Cool Dudes: The Denial of Climate Change Among Conservative White Males in the United States." *Global Environmental Change* 21, no. 4 (2011): 1163–72.

Miéville, China. "The Limits of Utopia." *Salvage*, no. 1 (2015): 177–89.

Moore, Jason W. *Capitalism in the Web of Life: Ecology and the Accumulation of Capital*. New York: Verso, 2015.

Nietzsche, Friedrich. *The Birth of Tragedy and Other Writings*. Translated by Ronald Spiers. Cambridge: Cambridge University Press, 1999.

Nietzsche, Friedrich. *Daybreak*. Translated by R. J. Hollingdale. Cambridge: Cambridge University Press, 1997.

Nietzsche, Friedrich. "The Dionysiac Worldview." In *The Birth of Tragedy and Other Writings*. Translated by Ronald Spiers. Cambridge: Cambridge University Press, 1999.

Nietzsche, Friedrich. *The Gay Science*. Translated by Josefine Nauckhoff. Cambridge: Cambridge University Press, 2001.

Nietzsche, Friedrich. "On Truth and Falsity in Their Ultramoral Sense." In *Early Greek Philosophy and Other Essays*. Translated by Maximilian A. Mügge. Edinburgh: T. N. Foulis, 1911.

Nietzsche, Friedrich. *Twilight of the Idols and The Anti-Christ*. Translated by R. J. Hollingdale. London: Penguin Books, 1990.

Norgaard, Kari Marie. *Living in Denial: Climate Change, Emotions, and Everyday Life*. Cambridge: The MIT Press, 2011.

Olson, Marie-Louise. "A Weather Man Breaks Down in Tears and Vows NEVER to Fly Again Due to Grim Climate-Change Report." *Daily Mail*, September 28, 2013. www.dailymail.co.uk/news/article-2436551/A-weatherman-breaks-tears-vows-NEVER-fly-grim-climate-change-report.html.

Pearce, Fred. *With Speed and Violence: Why Scientists Fear Tipping Points in Climate Change*. Boston: Beacon, 2007.

Pearson, Keith Ansell. "Living the Eternal Return as the Event: Nietzsche with Deleuze." *Journal of Nietzsche Studies*, no. 14 (1997): 64–97.

Samenow, Jason. "Meteorologist Eric Holthaus' Vow to Never to Fly Again Draws Praise, Criticism." *Capital Weather Gang* (blog), *Washington Post*, October 1, 2013. www.washingtonpost.com/blogs/capital-weather-gang/wp/2013/10/01/meteorologist-eric-holthaus-vow-to-never-to-fly-again-draws-praise-criticism/.

Smith, Daniel W. Translator's preface to *Nietzsche and the Vicious Circle*, by Pierre Klossowski, vii–xiii. Chicago: University of Chicago Press, 1997.

Vollmann, William T. *Carbon Ideologies*. 2 vols. New York: Viking, 2018.

Waite, Geoff. *Nietzsche's Corps/e*. Durham: Duke University Press, 1996.

4   Risk or security
Politics in the event

Carl Schmitt theorizes the event in the wake of Kierkegaard and Nietzsche. He does not expect God to intervene in the world but locates a similar interruptive force in sovereign political decisions. The sovereign becomes a secularized God and Kierkegaard's faith an act of obedience. But without a God who founds and maintains order, that task now falls to humans, who have no divine constraints on their actions.[1] For Schmitt, the event is a political exception in which socially organized affects become so intense that they upset political organization. While for Nietzsche, the instability of the event is a positive moment that opens experience to affects and experiences suppressed by routinized sociality, Schmitt favors arbitrary decisions made to homogenize social affect in the name of securing order. For him, this is a matter of exerting control to direct social forces which might otherwise lead to political transformations. This raises the question of what ethos one adopts toward such moments and whether one seeks to contain them or engage and propel the opening they create.

Schmitt has become one of the main thinkers used for thinking through the climate crisis. This is in part because climate change is feeding into the kinds of troubling political intensifications and situations that his thought dealt with, such as new kinds of exceptional violence, emergency measures to suppress environmental protest, the production of new exceptional spaces, and the possibility for the rise of dictatorial climate responses. A number of thinkers are working to resist Schmitt and the propensity his thought has to feed into injustice in such situations. But others like Bruno Latour have sought to rework Schmitt's thought for more productive responses to climate change. This chapter will attend to the dangerous side of Schmitt's thought while reading him against himself to suggest more emancipatory responses to political events. The critical appropriation pursued here gives a close reading of Schmitt, identifying what William E. Connolly calls a flashpoint "at which key existential investments enter the complex, sometimes unconsciously and sometimes as a juncture treated by the theorist as an undeniable starting point."[2] Through Schmitt's ethos of security, his theory became subservient to his personal politics. Against this, an ethos of risk opens more emancipatory responses to political events.

The chapter is divided into four sections. The first uses Schmitt to develop a theory of political events as intensifications within a system that are either subordinated to that order or break through to produce new political configura-tions. I then turn to critique Schmitt's analysis of political emergencies to argue for a partisan political position that can supplement the ways that Mann, Wainwright, and Latour have drawn on Schmitt in the context of climate change. The third sec-tion takes up Schmitt's notion of political theology and Catherine Keller's critique of it to examine the question of how to orient ourselves toward political events. The final section turns to Schmitt's existential ethos to argue that his commitment to security also shaped his vision of human nature, which in turn limited his abil-ity to attend to vital questions beyond the friend-enemy relation. Against Schmitt, I argue that an ethos of risk is necessary to motivate and sustain the partisanship necessary to politically respond to the existential condition of the Anthropocene.

The political event-machine[3]

In *Groundless Existence*, Michael Marder gives a productive reading of Schmitt's theory of the event. He begins by pointing out that for Schmitt the political does not operate in its own delimited sphere. As Schmitt argues, "The political can derive its energy from the most varied human endeavors, from the religious, eco-nomic, moral, and other antitheses. It does not describe its own substance, but only the intensity of an association or dissociation of human beings."[4] Marder sees in this an "unbounded versatility of the concept of the political," through which it transforms other domains. "Deprived of a playing field of its own, the concept reaches the heights of anti-foundationalism, as it shuns clear topographi-cal distinctions, transgresses ontological boundaries, and, as a result, acquires that plasticity which nourishes its ability to dwell in and to transmogrify all other domains."[5] Potentially inhabiting all other domains, the political manifests when tensions or antagonisms within one reach such a level of intensity that a relation of enmity emerges.[6] At this point, the original character of the domain is superseded by the political. A political event is the moment when the political instantiates itself within another domain, transforming it in the process.

Marder's insight is that Schmitt's concept of the political can help us attend to political transformations which eliminate old modes of being and develop new ones. "In the process of renewal that adumbrates the living connection between the form and the content of politics, expropriation facilitates the 'emergence' and 'formation' of new unities (forms of political existence) and functions as the inal-ienable aspect of decisions on the constitution as a whole."[7] While for Schmitt the political decision suspends the norm to preserve a way of life from a perceived threat, Marder turns that very suspension into a test of that way of life. For him, the groundlessness of the decision opens space for the emergence of new forms of political existence, despite the recalcitrance and inertia of outmoded forms. Marder thus anchors his reading of the event as transformation in the groundless-ness of the political: "The point of the political, like the point of the decision that lies at its core, is an instant of the greatest risk, an experience of groundlessness."[8] Here, however, Marder conflates risk and groundlessness. While this ambiguity allows him to develop a provocative conceptual reading of the political, it also yields a flat and potentially problematic understanding of the role that risk plays in effecting political transformation.

Marder initially distinguishes two types of risk, one taken by the sovereign, the other by the partisan. The former is "general risk," which applies to a population under threat and is readily mobilized for a conservative politics of security. The latter is the "pregnant risk" of the partisan, who engages in extralegal political action without the hope of security.[9] Marder later maintains this separation between sovereign and partisan risk,[10] but immediately after reaffirming this distinction, he claims that:

> In addition to the general theory of risk, the political configuration of decision-making calls for a more patient and meticulous 'risk analysis.' . . . The sundering apart of the process and the act replicates the divergence of means from ends in the gradations of partisan risk and performs Kierkegaard's 'leap of faith,' liberating itself from the discussions, deliberations, and calculations that prepare the ground for it. . . . The decision remains incalculable, subjective, absolute, and, therefore, risky in the pregnant sense of the term regardless of the field of meticulous calculations from which it takes off.[11]

Though general and pregnant risk are qualitatively different, Marder argues that the groundlessness of the decision, embodied here in Kierkegaard's leap of faith,[12] is rooted in pregnant risk, whether that decision is made by a sovereign or a partisan. The radical assumption of danger in consciously placing oneself outside a legal order to contest it is thus conflated with the general danger to a population that the head of the state wards of from a position of relative security.

This collapse of categories results from Marder reading the event in Schmitt through Derrida's *événement*. This is a notion of groundless expropriation that contains the reflexive autoimmunity to expropriate itself, in accordance with Schmitt's worry over the omnipresent possibility for the political to be depoliticized.[13] Marder's reading of Schmitt's concept of the political can be productively modified through Deleuze and Guattari's notion of the machine. The machine offers an alternative to Derrida's notion of *événement* that attends to both the plasticity and open-endedness of Marder's reading as well as the different kinds of risk that feed into either conservative or emancipatory transformations. It does so by analyzing political events through "displacement"[14] rather than groundlessness.

Like the political, a machine does not have a substance but describes a process. "To enter or leave the machine, to be in the machine, to walk around it, to approach it—these are all still components of the machine itself: these are states of desire, free of all interpretation. . . . Desire is not form, but a procedure, a process."[15] Not having its own substance, a machine is constituted by a constellation of desires in the social field and connections to other such machines of desire. Even if desires are not "inside" the political machine, they are connected to it. Individuals are part of the machine in their political and social investments, "but even more so in their adjacent activities, in their leisure, in their loves, in their protestations, in their indignations, and so on."[16] In Schmitt's words, "incessant friend-enemy disputes . . . embrace every sphere of human activity."[17] In this way, the waxing and waning of desire is folded into politics and the event.

An event occurs when desire "brings about a fully political and social invest-ment, engages with an entire social field."[18] As desires are added, subtracted, and intensified in the social milieu, machines are reconfigured, sometimes producing new political modes of being. The open-ended transformation that Marder identi-fies in Schmitt's concept of the political is found in the Kafka-machine as well but takes the form of displacement or "adjacency" rather than groundlessness.[19] Political transformation is a matter of moving to the next way of being or form of political organization.

By locating them within the larger machinic context, political events not only become sites of transformation but also resist Schmitt's enticement to make sov-ereign decisions self-sufficient. Thus, we might rethink how Schmitt's advocacy of the exception played out. Deleuze and Guattari write of K's desire for justice propelling him through the legal assemblage. Schmitt's judicial and philosophical commitments such as his interest in security also propelled him through the legal-political assemblage. In 1921, Schmitt's desire was machined into his argument about the use of emergency powers according to article 48 of the Weimar Consti-tution. The use of emergency powers under article 48 likely helped hold Weimar democracy together in the face of the social forces stressing it. Later, he argued that article 48 granted Hindenburg the opportunity to forestall coalescing Nazi political forces and to reassert the liberal regime. But these attempts to engage the exception may have helped reshape public desire, turning more people toward Hitler while simultaneously providing a legal backdrop that facilitated Hitler's production of a new fascist political form. This illustrates the complex working of desire in the machine: seeking to preserve the established order, Schmitt thought that emer-gency powers could be used to prevent Hitler's rise to power. He may not have paid enough attention to how their repeated invocation might weaken the Constitution and shift the public in Hitler's favor. It may also have prepared the public to accept the political exceptions invoked by Hitler. In a different way, the Left has recently tried to appropriate Schmitt's thought for its own purposes in ways he likely would not have foreseen. In yet another way, the Nazi Regime's political order likely intensified Schmitt's anti-Semitic desire in a way that the Weimar Republic did not. His anti-Semitism then became more prominent in his work. In yet another way, Schmitt's thought now enters into connection with nature and the hard sciences in the climate event. Thus, discontinuous displacement allows for a more open and contextualized reading of political events than that of absolute groundlessness.

The political transformations of machines vary. "Abstract machines" secure content through transcendental functions. In this instance, political events mobi-lize general risk to reinforce the existing system through conservative consolida-tions of power. Alternatively, concrete machines produce expressions that either give way to new machines or eventually get taken up within abstract machines. This encompasses those who accept pregnant risk to engage in an "immanent justice" in which they are "worth nothing except in themselves"[20] to produce "another possible community."[21] Yet Deleuze and Guattari indicate another way the transformative forces of the machine "measure the mode of existence . . . in terms of the capacity that [political assemblages] demonstrate for undoing their own segments."[22] Thus, it is no longer a matter of maintaining or transforming a way of existing but of the transformative potential contained in a given mode of existence. Such a machine has four characteristics: it has no need for transcendental law, it has the suppleness to realize itself immanently, it has the ability to immanently mobilize other social groupings, and it has the ability to organize those three characteristics to form this machine.[23] Such a machine draws on one aspect of Schmitt's concept of the political but pursues a line of flight from it. While Schmitt deals with events that interrupt order, he emphasizes responses that establish political stability. Deleuze and Guattari's machine reworks Schmitt's concept of the political to mobilize its interruptive potential for emancipatory politics, one form of which is the partisan.

Partisan politics

Reading Schmitt's pointed sentences, it is easy to reduce his thought on the political to a formulaic analysis of a strong leader declaring and battling an enemy. Yet as was argued in the previous section, the political can extend to all areas of life. This section will trace the decision through various instantiations of the political, expanding upon the generalization of it as determined by the leader of a regime. While Schmitt notes that decentralized political decisions are possible, his politics intervene in his theory to emphasize sovereign decisions that favor security. In doing way, he downplays partisan decisions for decentralized and potentially emancipatory political action. I argue that this is a flashpoint in his thought where he tries to limit who decides based on his own existential investments.

It is often claimed that Schmitt is concerned primarily with the state as the political actor.[24] While his language sometimes suggests that only the state can make a political decision, this exclusive reading cannot be maintained. When Schmitt speaks of the state as the political entity, he says that no defining criteria can be attached to the term "state" outside of it being the ultimate authority in the decisive case.[25] Schmitt further argues that all non-polemical definitions of the state are politically decisive, implying that either the definition itself or the person producing such a definition is politically motivated.[26] Third, Schmitt argues that the advent of the total state—when the state loses its monopoly, blends with civil society, and every issue becomes politicized—is politically decisive because it challenges state-centric politics.[27] Schmitt further argues that there are instances in which the political dissolves the state, either in the case of a civil war or in the possible emergence of a global state.[28] Finally, he seems to emphasize the role of the individual when he argues that: "Only the actual participants can correctly recognize, understand, and judge the concrete situation and settle the extreme case of conflict. Each participant is in a position to judge whether the adversary intends to negate his opponent's way of life and therefore must be repulsed or fought in order to preserve one's own form of existence."[29] There are many instances where Schmitt indicates that political decisiveness can be located outside of the state.

Schmitt's most significant extension of political decisiveness to non-state entities occurs in *Theory of the Partisan*. The partisan is defined by a relation of irregularity with regard to the state. Lenin brings this to the point where the state is explicitly set aside as a determining factor. "We must keep the development of the concept of the political in view, which precisely here takes a subversive turn . . . Since the onset of the 20th century, this state war with its bracketing has been destroyed and replaced by wars of revolutionary parties."[30] Even if the state was previously the model of the political, the 20th century sees it take second place to the revolutionary party.

Schmitt seeks to draw a line, however, when he addresses the "conceptual dissolution" of the partisan:

> In general, and in view of the rapid changes in the world, the tendency of traditional or 'classical' concepts, as one likes to call them today, to be changed or also to be dissolved is all too understandable . . . In a very important book for our subject the illegal resistance fighter and underground activist are made the true type of partisan . . . Illegality is substituted for irregularity; resistance, for military combat . . . Then, ultimately, any individualist and nonconformist can be called a partisan, without any consideration as to whether he would even think of taking up arms.[31]

Schmitt rejects the political transformation in which the political status of militant partisans is extended to those engaging in illegal resistance and civil disobedience, labeling it a "dissolution." He goes on to call upon states to impose criteria that would establish a clear international legal definition of the partisan based on militant irregularity. He thus reduces protest to criminality and excludes it from politics. This is a flashpoint in Schmitt's thought where his existential investments influence the politics of his thought.

Marder elides Schmitt's resistance to this dissolution, providing a selective citation of Schmitt's analysis in order to support his Derridean reading of the concept of the political:

> The process of concept dissolution that sees the partisan turn into everything and nothing in particular is most salient at a time of transition and, hence, in the emergence of a new *conceptual* unity. The minimal sense of ontological expropriation is epistemologically relevant to the concept of the political as well, permitting its form to adjust to the increasingly more significant partisan content and interspersing this period of adjustment with hyperbolic extensions and overvaluations of the partisan.[32]

But given that Schmitt resists this transition, Marder neglects a critical moment when the pregnant risk of the partisan is elided in favor of a state-centric politics of security. For Schmitt, definitions of what is political are themselves political. He sees this conceptual dissolution as a negative political development and thus intervenes against such a redefinition.[33]

Sitze clarifies the tension between Schmitt's analysis of political transformation and his political intervention. According to him, Schmitt attempts to think the representational crisis of modernity, which has no transcendent reference point with which to mediate and evaluate the world.[34] This has a particular effect on political analysis:

> The inconsistency of Schmittian science with itself—its permanent and constitutive openness to polemic, ideology, and propaganda—is utterly consistent with science in the Schmittian sense: it is the manifestation, in Schmitt's own criticism of modernity, of the crisis Schmitt thinks in and through his genealogy of the political, of his discovery that modern political institutions are radically and originarily incomplete in relation to their own attempts at peace, security, and reconciliation.[35]

Political thought that accepts the crisis of modernity as real is unable to move beyond the crisis to secure the validity of its thought and conclusions. It can thus easily slide from analysis into polemic. Schmitt acknowledges the transformation in political partisanship on the basis of his theory, even as he slides in the next sentences into criticizing it.

Schmitt and Sitze both refer to the tendency of modern political theory to become polemicism as "risk." Schmitt admits that political analysis is always subject to such risk.[36] Sitze criticizes him for succumbing to the risk of polemic in his Nazi politics, framing it as a self-betrayal expressive of careerism and opportunism.[37] Such a judgment, however, ignores risk in the pregnant sense. Marder, on the other hand, discusses pregnant risk but also ignores the problematic treatment that Schmitt gives it. Risk is not just limited to the potential for political thinkers to become polemicists, as with Schmitt's Nazism but extends to the personal risk of placing oneself outside the law. The activities of resistance that Schmitt excludes from partisan politics assume risk in this sense. At this flashpoint between risk and security, Schmitt resists the transition in partisan politics from irregularity to illegality, resistance, and non-conformism. At stake is the emergence of a form of political resistance that adopts a risky position in relation to the state and can destabilize society through its tendency toward progressive political transformation. Schmitt seeks to deny political legitimacy to such partisan politics, relegating them to criminality.

Slavoj Žižek identifies a similar move in Schmittian thought, which attempts "to depoliticize the conflict by bringing it to its extreme, via the direct militarization of politics . . . by reformulating it as a war between 'Us' and 'Them', our enemy, where there is no common ground for symbolic conflict."[38] Žižek sees symptoms of this militarization both in the radical Right recategorizing class struggle as class warfare and in the primacy that Schmitt accords to relations between sovereign states over the politics of social antagonism. In opposition to militarization, Žižek suggests that "a leftist position should insist on the unconditional primacy of the inherit antagonism as constitutive of the political" in which "the struggle for one's voice to be heard and recognized as the voice of a legitimate partner" involves "destabilizing the 'natural' functional order of relations in the social body."[39] It seems that this is the kind of social antagonism that Schmitt seeks to exclude when he resists the transformation of partisan politics. What Žižek overlooks, however, is that while Schmitt politically resists a more democratic partisanship, he recognizes its significance in his theory.

Sitze misses the emancipatory potential that pregnant risk expresses when social antagonism disrupts normal political processes. Epitomizing this, he provides a reading of Schmitt's *The Leviathan in the State Theory of Thomas Hobbes* to connect Schmitt's portrayal of Spinoza and Nazi ideology. In so doing, Sitze explains that Spinoza turns Hobbes on his head by putting forward "the extreme political position that it is not up to the sovereign to decide whether or not a given miracle is true, that this decision is instead up to me alone in my capacity as a reasoning and (possibly) pious being."[40] He even notes that it is for this reason that Schmitt views Spinoza as "the most radical and originary internal enemy of the modern state."[41] Sitze, however, concludes that Schmitt identifies Spinoza's thought with Jewish thought and thereby reduces Schmitt's rejection to anti-Semitism. But if Spinoza is the originary enemy, it is because he touches upon an element of the decision in Schmitt's theory, but which Schmitt pursues in a politically opposed direction. Spinoza was not only an embodiment the excluded, but the excluded play a positive role in his thinking. It is also the excluded who embody social antagonism when they disturb the normal order, and it is for this reason, in addition to anti-Semitism, that Schmitt rejects Spinoza. Ironically, Marder's theory of groundless political decisiveness leads him to side with Schmitt against Spinoza.[42] But the point at which the polemicization of political analysis meets the pregnant risk of a challenge to the political order must be acknowledged for Schmitt's theory to maintain its analytical force, even as he tries to suppress it.

For Sitze, Schmitt's failure in relation to his own project is his personal political engagement: "by turning his thought into the servant of his person, Schmitt in effect *arrested his own thought*, saving his proper name by suspending the very 'thought movement' that constituted the most forceful punch of his genealogical criticism of the modern."[43] But what Sitze sees as a character flaw is actually a political refusal on Schmitt's part to accept pregnant risk. Sitze is thus unintentionally literally correct when he points out that Schmitt's shortcoming is "a failure to come to terms with the possibility that the disintegration of the person he exposed in and through his own thought could . . . pronounce itself in his own person."[44] Schmitt was unwilling to risk his career or his personal security. Though Marder theoretically engages pregnant risk, he does not recognize Schmitt's political resistance to the transformative power it mobilizes when individuals expose themselves to crisis, persecution, imprisonment, and death. Interestingly, this is the point at which Sitze, Žižek, and Schmitt all converge: none of them follow the conceptual dissolution that leads to a new form of the political partisan because they all accept it, albeit in different ways, as a limit to Schmittian thought.[45]

Though they do not use the term, one provocative suggestion of how partisan pregnant risk contests Schmittean politics in the context of climate change comes from Joel Wainwright and Geoff Mann. Indeed, they have given perhaps the most thorough attempt to think through the political implications of Schmitt's thought in response to the event of climate change. They move past particular short-term emergencies that will provoke Schmittean responses to the large-scale developing climate emergency. Replicating Schmitt's language of stability vs. emergency, they point out that "A stable concept of the political can only hold in a relatively stable world environment; when the world is in upheaval, so too are the definition and content of the realm of human life we call 'political.'"[46] For them, the emergency is not one that manifests in any particular political system but changes the very foundation upon which those various legal and institutional systems exist thus requiring a change in the political itself.

Before moving to the political responses they map, it is necessary to look at their concept of political adaptation. They begin by critiquing the political adaptations that are already underway. The most prominent versions come from scientists and the IPCC and are often based on functionalist premises. The problem is that this obscures that "scientific *and* political decisions are at work" in such reports and proposals "in a way that makes these adaptations seem natural and functional."[47] This notion of adaptation, "rooted in the dominant philosophical and metaphysical traditions of the liberal capitalist global north," prescribes who will (or will not) pay, who will benefit, and who will suffer without engaging the political questions involved through reference to "apolitical" markets and scientific analyses.[48] Against this liberalism, Wainwright and Mann propose a more encompassing notion of the political.

> The political is not an arena in which dominant groups impose their interests and subaltern groups resist; it is, rather, the ground upon which the relation between the dominant and dominated is worked out. In other words, there is no nonpolitical or apolitical domination. Thus the fundamental adaptation that climate change demands of humanity is political in this sense.[49]

On the surface, this notion of the political seems to be in accord with Schmitt's. Yet they insist that, close as it is, it radically different in the inspiration it draws from Gramsci. While one key difference between the two—communism rather than fascism—comes down to personal politics, another has to do with scientific approach. While for Schmitt the exceptional moment breaks out of even its particular history, Gramsci's analysis is fundamentally historically rooted.[50] Wainwright and Mann make a strong point here, given that the history of the climate conjuncture is critical for any emergency decisions made it in or political adaptations made to it. Their analysis is thus more in line with the situational machinic displacement that I am arguing for than Marder's groundless expropriation.

Wainwright and Mann argue that the two critical axes that define the spectrum of political adaptations to climate change are capitalism and sovereignty. The first axis is critical because of the imbrications between capitalism, climate change, and justice, and the second because it defines the relation to the political exception brought on by climate change. On the basis of this, they map four Weberian ideal types of possible political futures: Climate Leviathan (capitalist planetary sovereignty), Climate Mao (non-capitalist planetary sovereignty), Climate Behemoth (capitalist anti-planetary sovereignty), and Climate X (non-capitalist anti-planetary sovereignty). I will deal with the first, because they think it the most likely, and the last, because it most directly takes up pregnant risk. Thus, these two possible trajectories address the two ends of the spectrum of Schmittean thought.

Climate Leviathan is a planetary regulatory authority "armed with democratic legitimacy, binding technical authority on scientific issues, and a panopticon-like capacity to monitor the vital granular elements of our emerging world: fresh water, carbon emissions, climate refugees, and so on."[51] Wainwright and Mann suggest that the United Nations Conference of Parties already serves as a nascent institutional form of this possibility and the 2015 Paris accord attests further to this. First, the proposals coming out of it have been thoroughly committed to capitalism and market-based solutions to climate change. Additionally, though they read Schmitt as a statist whose theory is in tension with this internationalism, ultimately it fulfills a more fundamental dimension of his thought. "It expresses a desire for, and the recognition of, the necessity of a planetary sovereign to seize command, declare an emergency, and bring order to the Earth, all in the name of saving life."[52] As I've argued above, while Schmitt was himself committed politically to the state, theoretically he was forced to recognize the significance of political decisions beyond the state such as the revolutionary party. Thus, there is little reason not to see this as an evolution of Schmittean thought in response to climate change. This possibility is worrying both in its pragmatic realism and the way it combines top-down authoritative decision-making that unjustly protects particular interests with the ideology and institutions of liberal democracy.

At the other end of the spectrum, Wainwright and Mann theorize a radical democratic response they call Climate X against planetary sovereignty and capitalism. Climate X is a cross-coalitional movement strongly informed by indigenous and anticolonial resistance as well as the anti-capitalist left, whose "priority must be to organize for a rapid reduction in carbon emissions" through a combination of techniques like mass boycott, divestment, strike, blockade, and reciprocity.[53] It sets itself in opposition to Schmitt in three key ways. First, it insists on "solidarity in composing a world of many worlds" as a key principle.[54] I think it could be argued that solidarity adheres to a Schmittean framework in taking up the friendship side of his friend-enemy distinction, but it is a side which he intellectually refused to develop and politically opposed in his preference for politics organized around enmity. Still, they do not discuss whether they effectually take up a politics of enmity against those seeking to continue capitalism through the expansion of planetary sovereignty. Rather, they argue that the friend-enemy distinction itself is illegitimate since it is based on assumed prior distinctions between "humans and nature . . . lives and life, humans and humanity . . ."[55] As we shall see in the final section, Schmitt would agree with them on this point: acceptance or refusal of his theory of the political comes down to ontological and existential commitments and assumptions. Second, against Schmitt's insistence on a homogenous political unity that can never encompass the entire world without falling back into friend–enemy divisions, Wainwright and Mann hold to "the inclusion and dignity of all" in such a way that "must not be homogenized into an ultimately universal-izing 'we.'"[56] They insist on a vision of global relations that are inclusive without being reductive. Finally, they refuse the category of sovereignty as such, even if for the moment it amounts to a "utopian gesture" against the so-called "realism" of Climate Leviathan.[57] But practically, it is the refusal of any politics that rules over a particular population and territory through the ability to declare which parts of the population will be saved and which not in an emergency. They argue that all sovereignty amounts to this and show that the priorities of Climate Leviathan, even in its inchoate form, would necessarily develop along these unjust lines.

On these three points, Wainwright and Mann productively think through how to navigate climate change without falling into the traps of Schmittean politics. Yet thinking with him a bit longer clarifies what it means to take up the anti-capitalist and anti-sovereign struggle of Climate X today. This question is critical given recent intensified violence against environmental journalists and activists, increasing attempts to codify environmental activism as terrorism, and active planning on the part of security agencies to employ exceptional measures to suppress protests that will arise in the face of environmental disasters. Even in rejecting Schmitt, one needs to stay with him to remain engaged with the extra-legal violence and suppression that will be used against the left by the state in a declared emergency. The partisans of Climate X need to take up the pregnant risk of the climate event and, far from embodying the secure decisiveness of the state, decide to expose themselves to the state's exceptional and perhaps proto-fascist legal measures in a struggle to transform that political system in favor of a more just world. This is a question of the ethos that will orient such movements toward emergency situations, which will be further explored in the final section.

***

Bruno Latour has given perhaps the most provocative reading of Schmitt's notion of the decision in relation to climate change. He argues that the New Climatic Regime[58] has created a new state of exception in which there is neither a sovereign political entity nor a set of natural laws that can serve as an arbiter to resolve the conflict over climate change. On the one hand, this means that we cannot rely on Science to discover and disseminate the universal laws of Nature to reach political consensus. On the other, it means we have to accept the risk of politics. "We shall never be able to repoliticize ecology without first agreeing to recognize that there is indeed a state of war . . . thanks to the disputes over the climate and how to govern it, we are asking the political question again in terms of life and death. What am I ready to defend? Whom am I ready to sacrifice?"[59] Some will be disturbed by the idea of giving up the universality of Science, others by the idea that this means political conflict with lives on the line, and yet others by Latour's reliance on Schmitt to work through this situation. Indeed, even though Haraway sympathizes with much of Latour's work, she argues that his engagement with Schmitt reduces his political approach to climate change to a matter of "trials of strength."[60] I think Latour reworks Schmitt in significant ways that go beyond this and thus it is worth examining why he turns to Schmitt and what possibilities it presents.

In addition to Schmitt's theory of politics, Latour draws on *Nomos of the Earth* for two reasons. First, Schmitt is helpful because he saw neither a unified nature nor particular disciplines as giving objective descriptions of it. Because of this, he does not see political territories as occupying neutral background space. "Space is the offspring of history . . . of multiple instances of territorialization, some of which would provisionally entail particular relations of *spacing*."[61] The process of producing territories also produces the land they occupy. For Latour, there is an important connection between the way space is produced as a result of history, technologies, and politics, and the way James Lovelock sees Gaia as produced by the organisms and material agencies that compose her. Thus, Schmitt helps to think about new ways of territorializing Earth.

The second reason is the way that Latour sees Schmitt orienting future conflict. There are three dimensions to this. First, future conflicts will no longer be extensive, but rather intensive. For Schmitt, this meant that the "discovery" of the new world that helped stabilize the European order was unrepeatable. Latour notes that while there were new conquests in the form of underground energy reserves, this era is over and there will be no extra-planetary territory to help resolve the conflict: it is limited to Earth. Second, though Schmitt's previous thought was more oriented toward war, here he argues that it a matter of those who realize they are at war but strive toward peace to take up the task of resolving conflicts. Latour expands this into an extensive framework of diplomacy. Third, Latour reads Schmitt's elaboration of the concept "nomos" as arguing for a "redistribution of agency," a process that he also pursues with the term "cosmogram," which he uses to "imagine the diplomatic assembly of the peoples struggling for the Earth."[62] For Latour, Schmitt not only helps us recognize the political task we have to take up after we recognize that we are in a state of war but also helps orient the ordering process through which that conflict might be resolved.

But Latour also finds it necessary to modify Schmitt's theory in some critical ways. He begins by refusing to reserve the decision for powerful leaders, arguing that "the political mode of existence is exceptional is *all its segments*."[63] Thus, he also affirms the decentralized political decision argued for above. Much more significant, however, is the way he turns Schmitt around on the question of distributing agency in the New Climatic Regime. "What Schmitt could not imagine was that the expression 'land-appropriation'—*Landnahmen*—could begin to mean '*appropriation by the land*'—that is, by the Earth."[64] What is at stake is the entry of Gaia into the political field. Far from being a neutral background or objective set of laws, Gaia appears as, among other things, a "threat" that multiplies "the sites in which radically foreign entities practice mutual 'existential negation.'"[65] Though Schmitt may be right about the state of war, he cannot see it as Latour does: engaged by a myriad of human and nonhuman agencies.

Given this understanding, Latour does not just ask people to define who their enemies and friends are. He calls for people to make a more extensive declaration that includes elements like what epoch they live in, what deity convokes them, what territory they seek to defend, what defines their people, and what distribution of agency (cosmogram) they adhere to.[66] For him, this new war of the worlds includes many different groups who can nonetheless roughly be divided into two sides: the Humans and the Earthbound. "Whereas Humans are defined as those who take the Earth, the Earthbound are *taken by it*."[67] The Humans live in the Holocene, in a unified but mute Nature known through universal Science. The Earthbound live in the Anthropocene and are aligned with those nonhuman agencies that keep emerging through the sciences in a world that is not unified but being composed.

In the Anthropocene, the friend–enemy lines and territories include methane, albedo, indigenous groups developing resilience strategies, rivers, wind farms, salmon, manufacturing processes, water vapor, urban plans, NGOs, modes of transportation, food systems, glaciers, belief systems, endangered animals, spreading diseases, and so on. "The territory of an agent is the series of other agents with which it has come to terms and that it cannot get along without if they are to survive in the long run."[68] The process of declaring territories makes clear the contours of the war currently under way. This declaration is necessary in order to begin engaging in diplomacy to reach a political resolution and a peaceful composition of the world.

There are a number of advantages to Latour's use of Schmitt. First, though it may seem like a provocation to violence, it is not clear that the business as usual scenario led by consumerism, global capital, nation-states, and militaries is any less violent.[69] Indeed, even leaving questions of inter-species violence to the side, the links between climate change and violence are already extensive and likely to become more so.[70] Thus, the declaration of war and drawing of territories may enable a more conscious, explicit, and honest acknowledgment and resolution of the forms of ecologically driven violence underway. Second, it allows ecologists to stop engaging an epistemological debate over the science and begin working for political action, calling for specific concessions from the opposing side. Third and perhaps most critically, it calls for shared sovereignty: "Gaia emerges not in order to reign in the place of all the States forced to submit to its laws, but *as that which requires that sovereignty be shared*."[71] This latter point not only radically reworks Schmitt's notion of sovereignty, which emphasizes its indivisibility, but also provides a framework for thinking more thoroughly about the kinds of human endeavors to pursue as collaborators rather than masters.

Yet one question remains. Given the problematic way that Schmittian decisions are influenced by existential dispositions toward either risk or security, is it wise to call on people to make such widespread decisions without attending more thoroughly to risk? Latour recognizes the risky element of diplomacy and politics and seems to acknowledge pregnant risk when pointing to "the state of exception in which [the Earthbound] agree to be placed by those whom they dare to call their enemies."[72] What is meant here is not clarified, but it at least suggests the danger of making an implicit war that is already being lost explicit. What would happen if many who vote for the center-left parties and support the Paris Climate Accord began demanding a much more radical political transformation, declaring the carbon states and carbon capitalism as enemies? What would happen if the national and corporate interests on the other side did not want to come to the table and engage in diplomacy?

Despite his acknowledgment of risk, it seems that Latour does not attend to this problem enough. When asking people to declare sides, it may be that many will decide in favor of short-term measures that secure only their own existence. Many people in privileged positions may prefer to have the airline industry, global supply chains, and border patrols as their allies rather than urban garden plots, resisting indigenous communities, radical climate scientists, and coral reefs. What happens if privileged carbon consumers accept without reflection Vinay Gupta's provocation that "the war has already begun and you did not notice because your side won?"[73] More troubling is that for Latour, the new political agencies of Gaia that have joined the battle equalize risk in a new way. With reference to those in France, California, New Guinea, Bangladesh, Beijing, and the territories of the Inuits, he declares: "The fact that all the collectives from now on . . . share the certainty that they fear nothing but 'being crushed by the falling sky' gives a totally different idea of universal solidarity."[74] He is right to note that climate change does open new lines for solidarity in the movement for ecological justice. But the risk is not distributed evenly. Many who join the frontlines may be put in perilous circumstances by losing just a little of what they have while others may refuse the call to solidarity even though they could lose much without significant suffering. Thus, the orientation that leads engaging pregnant risk needs to be attended to. The next section will rework Schmitt's political theology as a way of developing this orientation.

Thinking the event: political theology as approach

This section pursues a reading of *Political Theology* as a mode of thought oriented toward political events. It helps attend to the inevitable, risky, and existential character of such events, and the decisions they occasion. Schmitt inflects risk as an always possible threat, disregarding the role in plays in democratic struggle. In doing so, he strengthens the tendency to locate decisive power at the head of political regimes. Nonetheless, political theology as a mode of thought is open to a more decentralized approach to the event, as evidenced by Catherine Keller's theological reading of political theology in response to climate change.

Political theology is a "thought process," a "consistent thinking," and a "juristic thinking."[75] Schmitt uses the political-theological mode of thought to understand historical and contemporary orders, and the events that help constitute them.[76] Political theology uses the concept of sovereignty to focus on how power transforms order. "In political reality there is no irresistible highest or greatest power that operates according the certainty of natural law . . . The connection of actual power with the legally highest power is the fundamental problem of the concept of sovereignty."[77] Political-theological thought seeks to connect efficacious power to an established order. Sovereignty is not absolute power since no power can guarantee a specific order through time, nor can power be securely fitted to a specific position or person. On one hand, political theology seeks to designate the greatest coefficient of influence actualized through a decision. On the other, it must also consider the other powers that call that designation into question.

Though Schmitt focuses on the former, the latter is implied by the disconnect between actual and legal power.

For Schmitt, developing a mode of thought in order to analyze the event is necessary because events are inevitable points where actual power confronts legal power. An exception is an event in which an existing order is suspended and then reaffirmed or transformed. The event between orders is an open moment to which the preexisting norm cannot apply: "A transformation takes place every time . . . That constitutive, specific element of a decision is, from the perspective of the content of the underlying norm, new and alien. Looked at normatively, the decision emanates from nothingness."[78] Events call the existing order into question and provide opportunities for decisions to introduce something previously excluded, whether it is a normally illegal repressive measure or a new liberating form of political organization.

For this reason, events are risky. "The exception, which is not codified in the existing legal order, can at best be characterized as a case of extreme peril, a danger to the existence of the state, or the like. But it cannot be circumscribed factually and made to conform to a preformed law."[79] No system can anticipate the events that will challenge it, take it to its limits, and perhaps bring it to an end. Even in the midst of an event it is not clear how it will be resolved. But rather than describing the event as uncertain and risky, Schmitt makes it perilous, and for the state specifically. The risk of the event thus becomes something to be warded off, controlled, or combated, rather than accepted and engaged for the purpose of social transformation.

Even if the event is risky, Schmitt thinks that it should not be felt. When discussing forces that threatened to transform German society at the beginning of the 20th century, Schmitt criticizes "anxiety ridden" responses as failing to approach the situation through "cool-headed knowledge."[80] Such knowledge, gained through the political-theological approach, would realize that

> All new and great impulses, every revolution and reformation, every new elite originates from asceticism and voluntary or involuntary poverty (poverty meaning above all the renunciation of the security of the *status quo*) . . . Every genuine rebirth seeking to return to some original principle . . . appears as cultural or social nothingness to the comfort and ease of the existing *status quo*.[81]

Whether one seeks revolution or reformation, one must be willing to confront the risk of the event, which entails loss of comfort and a degree of asceticism, since once the exception is opened, the connections to the original or envisioned model are jeopardized. The danger involved in breaking with the given order is tied to the uncertainty of the outcome. Though Schmitt does not explicitly distinguish between the risk of trying to create something new through revolutionary activity and trying to reform a tight order, he does downplay the former. When he elucidates risk in the name of a "cool-headed knowledge" to overcome "anxiety," he excludes the terms of evental activism. In fact, the loss of comfort is only the first part of a politics of pregnant risk. While cool thought may be sufficient for leaders to consider the general risk to the population, the partisan requires a more engaged, affectively imbued mode of thought that draws on pregnant risk.

This matters for how one understands Schmitt's notion of "concrete life." "Precisely a philosophy of concrete life must not withdraw from the exception and the extreme case, but must be interested in it to the highest degree . . . The rule proves nothing; the exception proves everything."[82] The exception calls not only a particular order into question but also the ground on which it rests: the stability of the given mode of existence. The exception is an interruption that calls for a determination of public interest, state interest, safety, and public well-being.[83] In this way, the decision made in it seeks to secure the "everyday frame of life" to which rules and order are applied.[84] But this is a distant decision that is concrete only in the sense of general risk. The concrete situation of those who are immersed in the event and feel pregnant risk is different. Here, "concrete life" generates existential transformation in political subjects who seek to change the conditions of the routines, norms, politics, and assumptions of everyday life.

Yet the entire order need not come into question for an edge of evental openness to come into effect. Schmitt notes the "relativistic superiority" with which every jurist makes their assessments of normative laws and suggests that this produces minor events in his analysis of Kelsen's normative jurisprudence.[85] The same legal-political energy is tapped at the moment when a government creates a law and when an official enforces that law or decides not to, often applying it unevenly to different populations. Though Schmitt does not deal with them, it is clear that the latter minor suspensions can be extended to include each time a person engaged in civil disobedience calls a law into question. There are thus minor exceptions in which a decision is made that shapes the order without dissolving it.

Political-theological thought extends decisiveness to thoughts and perceptions concerning the legal order. "Such a decision in the broadest sense belongs to every legal perception. Every legal thought brings a legal idea, which in its purity can never become reality, into another aggregate condition and adds an element that cannot be derived either from the content of the legal idea or from the content of a general positive legal norm that is to be applied."[86] The decisive power of thought arises when it seeks to extend or challenge an established system: political theology occupies a place between legal ideals and existing law, awaiting actualization. The power to think beyond the dictates of a given order and to assess its potential to change provides a step toward reshaping that order. Critical thought is the first step of partisanship. The effect of each interpretation remains to be seen, but even those who are not functionaries of the state begin to alter the existing order by considering and advocating a different interpretation. Following this line, Marder argues that interpretation in Schmitt is laden with existential decisions: "Every interpretation is already an existential decision which is necessarily active, transformative, and reconstituting."[87] Interpretation expresses and reshapes desire in relation to the machine, the law, and the desires of others.

In a mode of deep partisan interpretation, Catherine Keller contests Schmitt's political theology to suggest a more radical and democratic response to climate change. For Keller climate change is a political and theological event that interrupts regular time with a kairotic moment. "The empty continuity of chrono-time is interrupted by a messianic contraction, in a decisive 'now' . . . a politico-spiritually charged transformation, immanent to the event of the Kairos."[88] The theological dimension helps to think this temporal exception and intensification. The political dimension calls attention to a human schema that may end and the existing public that has failed to do anything to prevent it. Indeed, it is expand-ing with increasing numbers of nonhumans caught up in the fate of the schema. Together, they create the urgency and exceptionalism that defines this event.

Keller contests Schmitt's reductive political theology as one in which theology stands in for a set of formal relations in secular politics in two ways. First, she opposes Schmitt's secularism to processes of secularization. The former is another "religion" in sense of being unhistorical, which seems to fit Schmitt's insistence that nothing can eliminate political theology as the primary problem of politics. Drawing on John Cobb, she argues for secularization as a process through which "religions" become attuned "to the challenge of social justice in the *saeculum*, the age and its politics."[89] Secularization materializes theology. Schmitt's insist-ence that political theology highlights the insufficiency of normative systems to address history is upset by a claim that a theory of politics based on a single moment of secularized theology faces historical insufficiency too, particularly in the climate event that defines this moment.

If a single secular social theory is not going to cut it, a more encompassing discussion of how the theological and the political intersect is necessary. There is "a largely hidden theology always at play, for good or ill, *within* the political" and that recognizing it opens "a vein of vivid transdisciplinarity in which theology itself may offer politically useful transcodings between the religions and the secu-lar."[90] Neither theology nor politics is reducible to the other. But neither are they self-sufficient. Nor is theology itself singular with one political message. For Kel-ler, political theology is a way "to think together with any who work to gather a de-essentialized, dense—indeed contracted—entanglement of our differences."[91] Theology gains ground in refusing universal secularism but also refusing any essential wholeness for itself, becoming political theology only in transdiscipli-nary work carried out with politics among other approaches.

Keller's welcome addition to efforts to respond to climate change goes beyond what I can discuss here, but I do want to pull out a couple of the implications of her reworking of political theology. Her pluralistic political theology critiques and develops Schmitt's notion of enmity. On one hand, Keller also rejects the "disempowering civility" of liberalism and affirms the importance of having ene-mies, even if just to name the homogeneous Schmittean practice of enmity as the enemy.[92] Then she brings in the theological dimension, suggesting that every politics requires a faith in the future against its uncertainty. The longer that faith remains unfulfilled, as it almost always will, the more that politics is driven to antagonism and enmity. Against this shortcoming, Keller allies her theological kairotic temporality to the politics of agonism developed by Mouffe and Con-nolly in their responses to Schmitt, as a way of attending to the complexity and uncertainty of a world in which bodies, peoples, and temporalities are entangled.[93]

On this basis, Keller redefines the political as "*collective assemblage across critical difference*. Critical difference signifies the crisis that difference effects for an emergent public, the divergence that demands fresh acts of self-organization."[94] Against Schmitt's always presumed unitary public defined by enmity, Keller develops the process of gathering public friendship in shared struggle as the criterion of the political. This public is never stable but emergent and always in a process of self-organization. The "we" of the public is never unitary and must maintain a connection with the undercommons, the dehumanized, and the nonhuman. This refigures the political event as always being on the edge of transformation: "the self-organization of a public will falsify any institution that obstructs it."[95] While Schmitt seeks to secure a regime against the indeterminacy of chronos, Keller proposes embodying the interruptive dimension of the kairotic political event.

The second main move that Keller makes against Schmitt is to oppose his exceptional moments of sovereign omnipotence with a notion of inception characterized by creative indeterminacy. She traces a line running from the logic of an omnipotent sovereign who decides on the exception to myriad injustices perpetuated by the exceptionalisms that flow from it such a nation, sex, race, and species. Against the relative certitude with which such measures are enacted, Keller's contestation comes from "negative theology, or *apophasis*" which is a "practice for liberating insight from certitude, for thinking at the edges of the unthinkable."[96] This is crucial not only given the complexities of climate change but also the complex becoming of a diverse public that would respond to it. Inflected with apophasis, the arbitrary sovereign decision becomes a decision that arises out of the "complex potentiality" of "the deep" or "chaos."[97] This decision amidst critical difference that no singular sovereign can think, name, or prescribe is what Keller calls inception, as opposed to Schmitt's exception. This is the perspective from which, even in this dark contracting moment of climate change, it is not sure that things will turn out badly, even if it is also not sure that they will come out well. "The inception signifies that present possibility—as active potentiality . . . It rumbles through the undercommons, it intersects multiply, it springs from the earth-ground of its time."[98] Here, theology, as a distinct but linked domain, pulls politics to attend to contracted time, emerging coalitions of beings, and transformative potential in a world whose unfolding remains unknown. Theology thus might "help now to gather a spreading public of shared Earth struggle."[99] Even when it comes to the pragmatic politics of assembling coalitions, Keller's theology calls us to attend to forces beyond the purview of Schmitt's narrow political theology. At the same time, for a position focused on politics rather than theology, inception can inform but cannot banish the exception. Indeed, the exception of pregnant risk, declared from below may be productive for engaging a kairotic event of becoming. The orientation of political theology has been developed, it is time to take up the ethos through which political events are engaged.

An ethos for political events

A political ethos orients not just how one responds to events but how they approach the world. This is why Mann and Wainwright point to it as a key factor motivating the various responses to climate change they seek to critique, such as Climate Leviathan and Climate Mao.[100] Critically, they do not specify an ethos motivating their own Climate X. Against the ethos pursued by Schmitt, this section will elaborate an ethos of pregnant risk that could contribute to that response.

Schmitt too is aware of how existential commitments shape an ethos that informs modes of analysis and perspectives on the world. One of his most fundamental commitments politicizes the world: all positions are political and any claim that a position is not political is also political. Even the most basic terms of analysis are adversarially politicized. "The polemical character determines the use of the word political regardless of whether the adversary is designated as nonpolitical (in the sense of harmless), or vice versa if one wants to disqualify or denounce him as political in order to portray oneself as nonpolitical (in the sense of purely scientific, purely moral, purely juristic, purely aesthetic, purely economic, or on the basis of similar purities) and thereby superior."[101] There are no fixed categories on which to base a stable analysis since all such terms are deployed in and inflected with politics. But such a view also reflects Schmitt's own priorities rather than establishing a neutral political analysis, which he admits, stating that it "presupposes a consistent and radical ideology."[102]

On the basis of this ideology, Schmitt holds to the inevitability of politicization: "Whether the extreme exception can be banished from the world is not a juristic question. Whether one has confidence and hope that it can be eliminated depends on philosophical, especially on philosophical-historical and metaphysical, convictions."[103] His analysis asserts that the exception is an inevitable part of any social order, but also that this is a matter of conviction rather than rational analysis. Another example concerns the possibility of a global community. Against the "normative ideal" that someday people will cease to group themselves according to the friend–enemy distinction, Schmitt holds to the "hope" and "pedagogic ideal" of the "inherent reality and the real possibility of such a distinction."[104]

That the political is inescapable is a conviction that both grounds his analysis and limits his vision. Schmitt tries to draw others into this vision, pushing readers to decide what they believe.[105] His strategy operates on two levels. First, Schmitt openly identifies points where a different set of investments would lead to different conclusions. Then, he mobilizes a vision of insecurity to push readers toward his politics of enmity. For example, he states that the definition of freedom is ultimately a matter of one's "anthropological profession of faith."[106] But then he suggests that freedom could have no meaning outside of a politics of coercion. This deceptive combination of apparent honesty and potential threat gives rise to concern about other ways of existing, inducing readers into worrying about securing their mode of human existence.

But as the discussion of Latour above showed, focusing on the tension between different human modes of existing elides how nonhuman processes affect human security, even as a singular species mode of existence is shown to be increasingly insufficient for handling the basic questions of quality of life, health, and security that Schmitt seeks to address. The pressure of politicization siphons attention and energy from the planetary dimensions that both feed into and exceed politics. This is why Latour seeks to broaden the Schmittian concern by bringing nonhuman actors into the political event.

Eliding a broader engagement with nature, Schmitt anchors his notion of political events in human nature: "One could test all theories of state and political ideas according to their anthropology and thereby classify these as to whether they consciously or unconsciously presuppose man to be by nature evil or by nature good . . . the problematic or unproblematic conception of man is decisive for the presupposition of every further political consideration."[107] The definitions of good and evil are not strict for Schmitt. Under evil he includes corruption, weakness, cowardice, stupidity, brutality, sensuality, vitality, irrationality, and so on. Under good he includes reasonableness, perfectibility, the capacity of being manipulated, of being taught, peacefulness, and others. In his vision, all of the theories that presume humans to be evil also argue that relations of enmity will always exist and that a decisive entity is necessary to produce temporary systems of relative stability and safety. Alternatively, theories presuming humans to be good assume that it is possible to eliminate enmity from the world and, therefore, that a stabilizing entity is not necessary. The former entails an inherent political aspect to the world, the latter that politics is not a fundamental part of life.[108] This is not merely a claim about human nature but an organization of that nature according to Schmitt's vision. While there are human characteristics in Schmitt's theory that speak to what might happen within the Anthropocene, he does not attend to how the human character brought on this era or what it means if the greatest threat is not an enemy.

The belief that humans are good is beyond Schmitt's system and can be accounted for within it only as a depoliticizing force. He thinks that humans are naturally evil, meaning that they are "by no means an unproblematic but a dangerous and dynamic being."[109] Even if humans achieve peace and stability for a time, it is in their nature to change and adopt new social configurations. For him, humans always have the potential to behave violently and "the entire life of a human being is a struggle and every human being symbolically a combatant."[110] Being alive involves some a struggle which is often carried into social relations. Here one sees his ethos starting to affect his analysis, since struggle is not necessarily combat and to see it as such is to unnecessarily militarize life while downplaying other vital aspects of political and existential struggle. Moreover, this position carries Schmitt further away from the ethos of Kierkegaard or Nietzsche, who both affirmed the world with its imperfections and difficulties.

Schmitt was too concerned with his own personal security to honestly analyze, let along commit, to political risk in the pregnant sense. Marder gives a good description of the ethos of pregnant risk, which

> cannot be manipulated because, in a state devoid of hope, the partisans do not anticipate a restoration of normalcy and security—at least not for themselves—and because they willingly assume the danger instead of evading it . . . . The partisan runs aground in the turbulence and uncertainty of the political, having refused to keep away from peril or to navigate around the

extremes in an attempt to negotiate a safe middle route between the threat of biological death and the certainty of political annihilation in an emergency situation . . . what instigates *partisan activity* is the total renunciation of rights in the spirit of *juridical passivity* and in an extreme reaction that overflows the distinction between the active and the passive subjective comportments, the Nietzschean dichotomy of self-affirmation and *ressentiment*.[111]

The hopelessness of this experience is not empty nihilism, because it refuses existing systems of legality, ethics, and rational behavior and instead draws its energy from the deep wells of possibility that contain both transformation and death. Self-affirmation and *ressentiment* both require existential coordinates within the social system, coordinates the partisan no longer maintains. Instead, pregnant risk is oriented toward future change. This suggests an emancipatory ethos for engaging and producing the event of political transformation.

This ethos is critical because of the ways that climate change will transform politics. Reactionary top-down interventions in the name of national security are already being planned in what Christian Parenti calls the "politics of the armed life boat" and more spontaneous authoritarian responses are likely as well.[112] Indeed, the rise of extreme right movements across the planet in the last decade, just as the effects of climate change are starting to be felt, makes such interventions more likely and raises the stakes for those pursuing alternative democratic responses. Imprisonment and murder of environmentalists is on the rise and the partisan nature of this activity is attested to in the fact that "environmental activists have higher death rates that some soldiers."[113] Even more troubling, a number of "democratic" regimes have begun to treat environmental protest as terrorism. But even the more benevolent green authoritarianism that some have argued is the only effective response to the climate and environmental crises is cause for concern.[114] The very efficacy that makes these top-down arguments tempting is what also exposes them to a Schmittean politics of security. As Mann and Wainwright stress, even in pursuing environmental goals, such decisive regimes are likely to preserve as much of the capitalist power structure as possible and exclude those who are already the most marginalized.

It is thus a critical moment to produce strategies for seizing the emergency from the bottom. This might entail the development of novel sovereignty sharing arrangements as in the case of the Yurok tribe who have agreed to risk a reduction in tribal sovereignty to work with the state of California as a way to increase space for biodiverse agencies in the region and, in doing so, extend tribal lands.[115] It also entails a more active engagement with the risks of traditional protest. Many environmental activists from DAPL Pipeline protesters to Tim DeChristopher to the myriad movements documented and promoted by Naomi Klein have accepted or even actively courted arrest. Perhaps no climate movement has made arrest such a central part of their strategy as Extinction Rebellion. In each of these cases, what comes forward is the politics of pregnant risk: intentionally placing oneself outside the security of an established legal order in an attempt to change it.

As the repressive responses of recalcitrant governments and their industrial allies intensify, the loss of security becomes greater. From the other side, the increasing impacts of climate change put additional existential pressure on marginalized and precarious peoples. Time is short, urgency is growing, risk is everywhere. In this reading of and against Schmitt, I have argued for the development of an ethos of pregnant risk first, as a way of transforming the defining boundaries of the existing political order and second, as a critical component for confronting and maintaining momentum against reactionary responses. Clive Hamilton gives a typical call for such a perspective: "We feel we owe obedience to a higher law even though we have to accept the consequences of disobeying the ones in the statute books."[116] That such consequences might include death is easily admitted but rarely openly discussed, even by radical leftists seeking to respond to climate change. But if killing is the extreme limit toward which Schmitt orients his politics of security, then death is the extreme limit toward which those seeking political transformation must orient themselves.

Notes

1　For more on this characteristic of modern conservatism, see Žižek, "Carl Schmitt in the Age of Post-Politics."
2　Connolly, *The Fragility of Things*, 99.
3　For a more detailed exploration of the argument made in this section, see Anfinson, "Risk or Security."
4　Schmitt, *Concept of the Political*, 38.
5　Marder, *Groundless Existence*, 62–3.
6　Though Schmitt defines the political relation more scientifically—"The distinction of friend and enemy denotes the utmost degree of intensity of a union or separation, of an association or dissociation" (Schmitt, *Concept of the Political*, 26)—his own investments in security lead him to focus exclusively on enmity. This refusal of political friendship, or solidarity, will be examined later.
7　Marder, *Groundless Existence*, 79.
8　Ibid., 67.
9　Ibid., 39–41.
10　Ibid., 54–5.
11　Ibid., 55.
12　Though metaphorically helpful, the comparison here is misleading. While for Kierkegaard, one can never establish the correctness of the decision, Schmitt believes that it can be established after the fact.
13　Marder, *Groundless Existence*, 61–2, 78–9.
14　Bensmaïa, "The Kafka Effect, xv–xvi.
15　Deleuze and Guattari, *Kafka*, 7–8.
16　Ibid., 81.
17　Schmitt, *The Leviathan in the State Theory of Thomas Hobbes*, 18.
18　Deleuze and Guattari, *Kafka*, 64.
19　Ibid., 7–8.
20　Ibid., 86.
21　Ibid., 17–18.
22　Ibid., 87.
23　Ibid., 85–8.

24 See, for example, Schwab's introduction to *The Concept of the Political*, 5–11, and Strong's forward to the same, pages xix–xx.
25 Schmitt, *Concept of the Political*, 19–20.
26 Ibid., 21.
27 Ibid., 22.
28 Ibid., 46–7.
29 Ibid., 27.
30 Schmitt, *Theory of the Partisan*, 49.
31 Ibid., 18–19.
32 Marder, *Groundless Existence*, 80.
33 For another instance where Schmitt makes a politically motivated argument against conceptual dissolution, see Anfinson, "Risk or Security."
34 Sitze, "A Farewell to Schmitt," 36–9.
35 Ibid., 45–6.
36 Ibid., 45.
37 Ibid., 46–50.
38 Žižek, "Carl Schmitt in the Age of Post-Politics," 29.
39 Ibid., 28–9.
40 Sitze, "A Farewell to Schmitt," 52.
41 Ibid.
42 Marder, *Groundless Existence*, 77–8.
43 Sitze, "A Farewell to Schmitt," 49.
44 Ibid.
45 Ibid., 30, 64–6.
46 Wainwright and Mann, *Climate Leviathan*, xi.
47 Ibid., 67, 71.
48 Ibid., 72–8.
49 Ibid., 80.
50 Ibid., 87.
51 Ibid., 30.
52 Ibid., 31.
53 Ibid., 173, 189–90, 197.
54 Ibid., 176.
55 Ibid., 177.
56 Ibid., 176, 188.
57 Ibid., 192.
58 This term is connected but not reducible to the "Anthropocene."
59 Latour, *Facing Gaia*, 227.
60 Haraway, *Staying with the Trouble*, 41–3.
61 Latour, *Facing Gaia*, 231–2.
62 Ibid., 235.
63 Ibid., 229.
64 Ibid., 251.
65 Ibid., 238, 244.
66 Ibid., 151.
67 Ibid., 251.
68 Ibid., 252.
69 Ibid., 239–41.
70 Climate influenced violence takes many forms. See for example: Mann, *The Hockey Stick and the Climate Wars*; Mitchell, *Carbon Democracy*; Welzer, *Climate Wars*; Parenti, *Tropic of Chaos*.
71 Latour, *Facing Gaia*, 280.
72 Ibid., 251.

73 Gupta, "Time to Stop Pretending."
74 Latour, *Facing Gaia*, 216.
75 Schmitt, *Political Theology*, 1–2, 46, 48.
76 Ibid., 2.
77 Ibid., 17–8.
78 Ibid., 31–2.
79 Ibid., 6.
80 Schmitt, "The Age of Neutralizations and Depoliticizations," 93–4.
81 Ibid., 94.
82 Schmitt, *Political Theology*, 15. Schmitt repurposes this insight from Kierkegaard.
83 Ibid., 6.
84 Ibid., 13.
85 Ibid., 20–1.
86 Ibid., 30.
87 Marder, *Groundless Existence*, 8–9.
88 Keller, *Political Theology of the Earth*, 4.
89 Ibid., 12–13.
90 Ibid., 10.
91 Ibid., 11.
92 Ibid., 24–6.
93 Ibid., 28.
94 Ibid., 33.
95 Ibid.
96 Ibid., 15.
97 Ibid., 42–3.
98 Ibid., 59.
99 Ibid., 66.
100 Mann and Wainwright, *Climate Leviathan*, 58.
101 Schmitt, *Concept of the Political*, 30–2.
102 Schmitt, *Political Theology*, 42.
103 Ibid., 7.
104 Schmitt, *Concept of the Political*, 28.
105 Ibid., 53–4.
106 Ibid., 57–8.
107 Ibid., 58.
108 Ibid., 58–68.
109 Ibid., 61.
110 Ibid., 33.
111 Marder, *Groundless Existence*, 39–40.
112 Parenti, *Tropic of Chaos*, 11.
113 Nuwer, "Environmental Activists Have Higher Death Rates Than Some Soldiers."
114 One can read the Climate Leviathan and Climate Mao responses in this way, but others have actively advocated for it. See Orr, *Down to the Wire*; Mckibben, "We Need to Literally Declare War on Climate Change"; Parenti, "Reading Hamilton from the Left."
115 Kormann, "How Carbon Trading Became a Way of Life for California's Yurok Tribe."
116 Hamilton, *Requiem for a Species*, 225–6.

Bibliography

Anfinson, Kellan. "Risk or Security: Carl Schmitt's Ethos of the Event." *Telos* 181 (Winter 2017): 85–105. doi:10.3817/1217181085.

Bensmaïa, Réda. Forward to *Kafka: Toward a Minor Literature,* by Gilles Deleuze and Félix Guattari, ix–xxi. Translated by Terry Cochran. Minneapolis: University of Minnesota Press, 1986.

Connolly, William E. *The Fragility of Things: Self-Organizing Processes, Neoliberal Fantasies, and Democratic Activism.* Durham: Duke University Press, 2013.

Deleuze, Gilles, and Felix Guattari. *Kafka: Toward a Minor Literature.* Translated by Dana Polan. Minneapolis: University of Minnesota Press, 1986.

Gupta, Vinay. "Time to Stop Pretending." Talk Given at Uncivilization Dark Mountain Festival, Llangollen, Wales, May 29, 2010. www.youtube.com/watch?v=EkQCy-UrLYw.

Hamilton, Clive. *Requiem for a Species.* New York: Earthscan, 2010.

Haraway, Donna. *Staying with the Trouble: Making Kin in the Chthulucene.* Durham: Duke University Press, 2016.

Keller, Catherine. *Political Theology of the Earth: Our Planetary Emergency and the Struggle for a New Public.* New York: Columbia University Press, 2018.

Kormann, Carolyn. "How Carbon Trading Became a Way of Life for California's Yurok Tribe." *The New Yorker,* October 10, 2018. www.newyorker.com/news/dispatch/how-carbon-trading-became-a-way-of-life-for-californias-yurok-tribe.

Latour, Bruno. *Facing Gaia: Eight Lectures on the New Climatic Regime.* Translated by Catherine Porter. Medford, MA: Polity Press, 2017.

Mann, Michael. *The Hockey Stick and the Climate Wars: Dispatches from the Front Lines.* New York: Columbia University Press, 2014.

Marder, Michael. *Groundless Existence: The Political Ontology of Carl Schmitt.* New York: Continuum, 2010.

Mckibben, Bill. "We need to Literally Declare War on Climate Change." *The New Republic,* August 15, 2016. https://newrepublic.com/article/135684/declare-war-climate-change-mobilize-wwii.

Mitchell, Timothy. *Carbon Democracy: Political Power in the Age of Oil.* Brooklyn: Verso, 2011.

Nuwer, Rachel. "Environmental Activists Have Higher Death Rates Than Some Soldiers." *Scientific American,* August 5, 2019. www.scientificamerican.com/article/environmental-activists-have-higher-death-rates-than-some-soldiers/.

Orr, David W. *Down to the Wire: Confronting Climate Collapse.* New York: Oxford University Press, 2009.

Parenti, Christian. "Reading Hamilton from the Left." *Jacobin,* August 26, 2014. www.jacobinmag.com/2014/08/reading-hamilton-from-the-left/.

Parenti, Christian. *Tropic of Chaos: Climate Change and the New Geography of Violence.* New York: Nation Books, 2011.

Schmitt, Carl. "The Age of Neutralizations and Depoliticizations." In *The Concept of the Political.* Translated by Matthias Konzen and John P. McCormick. Chicago: University of Chicago Press, 1993.

Schmitt, Carl. *The Concept of the Political.* Translated by George Schwab. Chicago: University of Chicago Press, 2007.

Schmitt, Carl. *The Leviathan in the State Theory of Thomas Hobbes.* Translated by George Schwab and Erna Hilfstein. Chicago: University of Chicago Press, 1996.

Schmitt, Carl. *Political Theology.* Translated by George Schwab. Chicago: University of Chicago Press, 1985.

Schmitt, Carl. *Theory of the Partisan.* Translated by G. L. Ulman. New York: Telos Press, 2007.

Schwab, George. Introduction to *The Concept of the Political*, by Carl Schmitt, 3–16. Chicago: University of Chicago Press, 2007.

Sitze, Adam. "A Farewell to Schmitt: Notes on the Work of Carlo Galli." *New Centennial Review* 10, no. 2 (2011): 27–72.

Strong, Tracy. Forward to *The Concept of the Political*, by Carl Schmitt, ix–xxxi. Chicago: University of Chicago Press, 2007.

Wainwright, Joel, and Geoff Mann. *Climate Leviathan: A Political Theory of Our Planetary Future*. New York: Verso, 2020.

Welzer, Harald. *Climate Wars: Why People Will Be Killed in the Twenty-First Century*. Translated by Patrick Camiller. Malden, MA: Polity Press, 2012.

Žižek, Slavoj. "Carl Schmitt in the Age of Post-Politics." In *The Challenge of Carl Schmitt*, edited by Chantal Mouffe. London: Verso, 1999.

5 Power and possibility
Techniques of the event

> Had Foucault entertained ontological intentions, he could have indeed claimed
> that all truthful Being is of the nature of lightning. The meaning of Being is not
> existence and the timeless preservation of essence, but event, the opening up of
> the horizon, and the spawning of temporary orders . . . Foucault accomplished
> the breakthrough to a foundational research oriented toward Event philoso-
> phy . . . [and] stepped up to the challenge of rethinking the core of all philoso-
> phy, the theory of freedom: no longer in the style of a philosophical theology of
> liberation—also known as alienation theory, but as a doctrine of the Event that
> liberates the individual and in which he moulds and risks himself.[1]

In the previous chapters, Kierkegaard pushed the insufficient individual to be
existentially transformed through the event, Nietzsche connected events across
the individual, social, and planetary scales, and Schmitt drew our attention to the
implications of political events rumbling underneath and sometimes erupting in
political orders. As Sloterdijk's assessment in the quote above suggests, Foucault
combines each of these dimensions in his own approach to the event. His orientation toward the collective history of Western knowledge pushed him to explore,
define, and transgress its limits. To do this he developed a method of explanation
in which he refused to adopt already given, unitary, necessary, inevitable, or extra-
historical modes of analysis. He sought to analyze the event as a problem rather
than a given and, in so doing, to implicate the researcher in the process of knowledge production. To bring out the elusive double-edge of these events, Foucault
produced a variety of concepts, techniques, descriptions, and experiences to shape
both knowledge and those producing it.

This approach may serve as a critical point of entry in debates around climate
change, where knowledge about the problem seems to be largely disconnected
from the power to do anything about it. Naomi Oreskes and Eric Conway have
diagnosed this disconnection as a failure rooted in the false assumption of Baco-
nian philosophy that scientific knowledge about the world translates into power.
The failure to respond to climate change despite overwhelming evidence of what
causes it and the harmful effects it produces shows how deep this disconnection
can be.[2] This is disheartening for many, particularly those whose hope and strategy

is to insist on scientific knowledge as the path to forging a political response. But the disconnect is intensified by the way a neoliberal theory built to protect distributed individual freedom concentrates and secures political power for a small economic elite.[3] This is why Bruno Latour has suggested that capitalism has unevenly distributed the power to respond to climate change.[4]

Foucault helps us to be neither surprised nor disheartened by these developments. His method of "eventalization" has two "theoretico-political functions."[5] First, "it means making visible a *singularity* at places where there is a temptation to invoke a historical constant, an immediate anthropological trait, or an obviousness which imposes itself uniformly on all. To show that things 'weren't as necessary as all that.'"[6] The inability of climate science to direct action does not indicate the collapse of either true knowledge or the essence of humans as rational creatures. Rather, it suggests the emergence of a particular moment and way of being. It is thus critical to ask what conditions made its emergence possible. This is the second function of eventalization, which entails "rediscovering the connections, encounters, supports, blockages, plays of forces, strategies and so on which at a given moment establish what subsequently counts as being self-evident, universal and necessary."[7] Focusing on the event pushes Foucault away from an analysis of causality that produces final explanations and toward a notion of intelligibility that draws out the contingent characteristics and processes that constitute an event. Eventalization thus reveals latent connections of power and forges new possibilities between the context of the event and the researcher's own.

Far from relativizing scientific truth, this mode of questioning is an ethos of the event that empowers the subject. This is what Foucault means when he says that eventalization goes beyond method to express a political spirituality.

> 'What is history, given there is continually being produced within it a separation of true and false' . . . How can one analyze the connection between ways of distinguishing true and false and ways of governing oneself and others? The search for a new foundation for each of these practices, in itself and relative to the other, the will to discover a different way of governing oneself through a different way of dividing up true and false—this is what I would call 'political *spiritualité*'.[8]

Foucault connects a way of doing history, a way of examining one's own presuppositions and a way of governing oneself and others differently through the event. If Foucault was able to find new forms of freedom amidst penal systems and neoliberal governmentalities, then perhaps the situation with regard to climate change today is not as dire as Oreskes and Conway suggest. Climate change is thus not a failure of reason, as Jamieson argued in Chapter 1. Rather, the geohistorical event of the Anthropocene manifests as a novel set of possibilities: many of them are dangerous but others are liberating. The ethos of the event developed by Foucault for understanding limitations, possibilities, and modes of action is critical for empowering responses to climate change.

In the first part, I will address Foucault's notion of a "limit-experience." Limit-experiences produce an unlivable yet insightful tension in the self that opens the possibility for personal and political reconfiguration. The second part looks at how events occur at multiple levels while sharing degrees of connection, such as that between limit-experiences as micro-events and transitions between different *epistemes* as macro-events. Our current *episteme* is characterized by the figure of the human, yet the Anthropocene might inaugurate a new episteme. In the third section, I turn to Foucault's work on care of the self. Caring for oneself is a matter of developing an ethos that instills an active relation to the event within oneself, equipping the subject to respond to unexpected events and helping align personal conduct with the truths to which it holds. Such an ethos can play a critical role in responding to climate change. The fourth section uses the case of environmentalist Paul Kingsnorth to suggest what a Foucauldian ethos of the event might look like in the Anthropocene. What emerges is an active politics of transformation that is critical for matching the shifts and uncertainties entailed by climate change.

Limit-experience as event

The idea of a limit-experience was crucial for Foucault, yet he wrote little about it. His most extensive discussion of the concept occurs in a 1978 interview with Duccio Trombadori. In the interview, Foucault claims that "The idea of a limit-experience that wrenches the subject from itself is . . . what explains the fact that however boring, however erudite my books may be, I've always conceived of them as direct experiences aimed at pulling myself free of myself, at preventing me from being the same."[9] The most important characteristic of the limit-experience for Foucault is its transformative potential, which is carried forward in a variety of different ways throughout Foucault's work. A limit-experience incites transformation by making experience intense to the point of being unlivable. "What is required is the maximum of intensity and the maximum of impossibility at the same time . . . experience has the function of wrenching the subject from itself, of seeing to it that the subject is no longer itself, or that it is brought to its annihilation or its dissolution."[10] Limit-experiences are uncontrollable and do not lead to a predictable outcome. Rather, the transformation is propelled by an experience so intense that one must become otherwise, without knowing what direction will be taken. The subject becomes a malleable object of the experience.

Yet it is not a matter of passively waiting for such experiences to sweep over the subject. Without being able to intentionally create a limit-experience, Foucault nonetheless suggests that one can set the stage for them. "The experience through which we grasp the intelligibility of certain mechanisms (for example, prison, punishment, and so on) and the way in which we are enabled to detach ourselves from them by perceiving them differently will be, at best, one and the same thing."[11] Foucault tries to increase the intensity of experience by making knowledge felt. It is not the abstract intelligibility of reason, but a felt intelligibility. A limit-experience is an event in which once we see how things are, we must live otherwise than we have. But as Foucault notes, this forcing is also an enabling: we become capable of freeing ourselves from a problematic situation in which we might otherwise have remained.

Foucault thus engages limit-experiences both for the negative moment when they tear us away from ourselves as well as for the way they enable us to construct ourselves.[12] He engaged that process of construction through the books he wrote. As Foucault states, "I'm an experimenter in the sense that I write in order to change myself and in order not to think the same thing as before."[13] This transformation goes beyond the personal to attempt to reconstruct the political as well. Foucault's limit-experiences were transmuted into public interventions through his work and activism. "In the book, the relationship with the experience should make possible a transformation, a metamorphosis, that is not just mine but can have a certain value, a certain accessibility for others . . . this experience must be capable of being linked in some measure to a collective practice, to a way of thinking."[14] Since the individual is enmeshed in a collective milieu, events in the individual can ripple outward and be magnified through public engagements such as books and even further amplified if they resonate with related social interventions. Thus, Foucault engaged experiences that pressed upon him and brought people into contact with a piece of the world such that they had a different experience of it.

In this way, he tried to open the possibility for others to have limit-experiences by using history to interrupt the present, revealing its structures and thereby inciting transformation. The goal is "to invite others to share an experience of what we are, not only our past but also our present, an experience of our modernity in such a way that we might come out of it transformed."[15] Though Foucault's work is historical, he makes it intrude into the present as a clarifying moment. He argues that the effect of this technique can be seen in the reaction to *Madness and Civilization*.

> The book stops at the very start of the nineteenth century . . . Despite all this, the book has continued to figure in the public mind as being an attack on contemporary psychiatry. Why? Because for me—and for those who read it and used it—the book constituted a transformation in the historical, theoretical and moral or ethical relationship we have with madness, the mentally ill, the psychiatric institution, and the very truth of psychiatric discourse.[16]

Foucault did not write about contemporary psychiatry, but rather discussed some historical aspects of psychiatry in such a way that people in the present understood their situation differently. If the text itself does not describe the present then it is not new knowledge that clarifies the situation, but a new degree of felt intelligibility. History mixed with the present to become an event that set other reactions and transformations in motion.

The individual and collective transformative potential of such limit-experiences can be seen more extensively in the reaction of people connected with the prison industry to *Discipline and Punish*.

> When the book came out, different readers—in particular correctional officers, social workers, and so on—delivered this peculiar judgment. 'The book is

paralyzing. It may contain some correct observations, but even so it has clear limits, because it impedes us; it prevents us from going on with our activity.' My reply is that this very reaction proves that the work was successful, that it functioned just as I intended. It shows that people read it as an experience that changed them, that it prevented them from always being the same.[17]

Here, a limit-experience rippled through specific sectors of the social fabric. Provoked by his own experiences, Foucault composed a history of the prison. This had effects upon academics as well as helped propel Foucault's political activities. The effects of this experience reached even farther, into the public and those working in connection to the prison system. The impact of a new degree of intelligibility on daily work activities highlights the intensity of these limit-experiences while showing their potential to spur collective action.

Yet the worth, particularly with regard to politics, of Foucault's theory of limit-experiences has been called into question. Martin Jay critically defends the relevance of limit-experiences, arguing for their value against both those who seek to use "experience" as a secure foundation for the individual subject and those who dismiss experience as solely a matter of discourse. In particular, he focuses on the criticism of limit-experiences given by James Miller in his biography of Foucault. Miller argues that if one looks at both Foucault's life and work, one ends up seeing a single consistent self throughout. For Jay, this misreading "works to smooth over the palpable tensions, even contradictions, that make the concept of limit-experience so productive and fascinating."[18] Keeping the productive tensions intact, Jay argues that "[w]hat Foucault seems to mean by limit-experience, then, is a curiously contradictory mixture of self-expansion and self-annihilation, immediate, proactive spontaneity and fictional retrospection, personal inwardness and communal interaction."[19] This description highlights two important aspects neglected by Miller. First, instead of combining both positive and negative experiences under the unity of a single consistent self, Jay highlights the change that occurs between one self and the other. Second, though Miller does not give much credence to the collective aspect of limit-experiences, it is crucial for Foucault.

For Jay, however, Foucault's point is that experience is plural and that no single version can serve as a foundation for truth or politics.[20] This is emphasized by the critical questions with which Jay marks "the limits of limit-experience" to keep it from doing more than being one more kind of experience among others. But Foucault goes further to argue that an approach to the world that embraces limit-experiences is more appropriate to the kind of beings we are than one that avoids them. He also thinks that in a society with extensive systems of knowledge and control, limit-experiences can be politically critical. But beyond these points, Foucault thinks that limit-experiences take the place of philosophical truth in the modern historical era.

It was an event when we entered the era defined by the limit and the continual transgression of it.[21] The death of God characterizes this era.

> Not that this death should be understood as the end of his historical reign, or as the finally delivered judgment of his nonexistence, but as the now-constant

space of our experience. By denying us the limit of the Limitless, the death of God leads to an experience in which nothing may again announce the exteriority of being, and consequently to an experience that is interior and sovereign. But such an experience, for which the death of God is an explosive reality, discloses as its own secret and clarification, its intrinsic finitude, the limitless reign of the Limit, and the emptiness of those excesses in which it spends itself and where it is found wanting. In this sense, the inner experience is, throughout, an experience of the impossible.[22]

The death of God is the "now-constant space of our experience." Implicit in this explanation is Nietzsche's famous section on the Madman in *The Gay Science*. Nietzsche describes the death of God there as a continual falling—a grasping and missing without orientation.[23] The jabs of the atheists in that aphorism, just as those of contemporary "new atheists," amount to nothing more than the banal transgression of customs and manners. Contemporary sincere professions of belief in God are seen as more transgressive, which is part of the force of Kierkegaard's thought. And in Nietzsche's aphorism, even the madman's attempt to take the death of God seriously is a transgression: the authorities haul him away. Each attempt to secure experience ends up being excessive, comedic, vulgar, or mad, underscoring the fact that we no longer live under "the limit of the Limitless," but rather "the limitless reign of the Limit." Such attempts and experiences are not, for all that, either equivalent or pointless. That this form of experience is now interior and sovereign means that finitude is now confronted in various guises. Though we cannot continuously live at the limit, our experience is constituted by occasionally passing through its limits. If this contradictory experience is "impossible," then it is this characteristic that Foucault orients himself toward and intensifies in a limit-experience.

More concretely, "the death of God restores us not to a limited and positivistic world but to a world exposed by the experience of its limits, made and unmade by that excess which transgresses it."[24] Just as removing sexuality from a Christian context and placing it within scientific or erotic discourses does not return eroticism to its true nature, so the world is not made natural by the death of God. We do not experience it as a limited place confined to positivist explanations and modes of being. Our entry into the Anthropocene only emphasizes this more: the unveiling of heretofore-unknown planetary limits and the limits of positivist knowledge to describe the causes, effects, and solutions of this "natural" phenomenon. Limit-experiences are important because they provide a sense of the role of the event in life: the making and unmaking of the world and the refashioning of the limits that compose it.

Foucault seeks to gain a sense of this process and even to shape it through transgression. "Transgression carries the limit right to the limit of its being; transgression forces the limit to face the fact of its imminent disappearance, to find itself in what it excludes . . . to experience its positive truth in its downward fall."[25] The limits that constitute experience are forms of exclusion. They exclude peoples, cultures, other ways of being, other modes of thought and so on, but by this very exclusion they shape experience. The high value the West places on its standard of life is tied to its limited ability to take seriously the potential that climate change has to end that standard of life. The idea of infinite growth forms another limit, making modes of life beyond it unimaginable. Transgression is a way not only of realizing how the limit owes its being to this exclusion but also of experiencing something that is excluded. Limits and exclusions are thus reworked through transgression. What interests Foucault in transgression is the transformation that takes place through it. He fashioned an ethic out of transgression called "nonpositive affirmation."

> Transgression contains nothing negative, but affirms limited being—affirms the limitlessness into which it leaps as it opens this zone to existence for the first time. But, correspondingly, this affirmation contains nothing positive: no content can bind it, since, by definition, no limit can possibly restrict it. Perhaps it is simply an affirmation of division . . . only retaining that in it which may designate the existence of difference.[26]

Nonpositive affirmation affirms the limit in three ways: it affirms the uncertain and insecure act of transgressing the limit, it affirms a world built and continuously rebuilt upon contingent limits, and it affirms limited being, an existence which develops only through transgressing one limited mode of being to move to the next. Because being is limited, no position can be secured or made transcendent.

The ethic of nonpositive affirmation can best carried out through a mode of being that actively engages limit-experiences. Foucault develops this mode of being, seeking "an experience that has the power 'to implicate (and to question) everything without possible respite,'" since it founds "a philosophy which questions itself upon the existence of the limit."[27] A philosophy of the limit-experience reworks its own foundations. It does so out of a desire to experience that which is excluded, which is also that which is possible. Its primary orientation is to see what kinds of existence are possible by reworking the limits, divisions, and markers of difference that structure life. Thus, it interrogates itself at the place from which the world receives its consistency, pursuing what Foucault saw as Nietzsche's project of producing "absolutely and in the same motion, a Critique and an Ontology."[28] Transgression is a philosophy, an experience, and a way of life that undertakes critique through limit-experiences to transform the subject and the world.

For this reason, limit-experiences are more important to Foucault than Martin Jay makes them out to be. For the latter they are simply one way among others of questioning both transcendental and discursive takes on subjectivity and experience. Foucault, however, sees them as a particularly critical mode of transforming the self and society in the contemporary era. When the modes of being that give self and society stability are composed around specific limits, it is transgression which both brings us closest to the "truth" of our era and gives us the greatest facility for moving through it and experiencing its different possibilities. Yet for opening so much possibility, limit-experiences remain elusive and unruly. They cannot be willed and even when one does emerge, the subject comes to terms with them only after the fact through an experimental process of construction. To develop a more encompassing understanding of the event, Foucault reworked his analysis of the limit-experience by examining it in relation to other transformations.

Connecting micro- and macro-events across scale

Though it examines transformations on scales beyond the personal, Foucault's *The Order of Things* began as a limit-experience.

> This book first arose out of a passage in Borges, out of the laughter that shattered, as I read the passage, all the familiar landmarks of my thought—*our* thought, the thought that bears the stamp of our age and our geography—breaking up all the ordered surfaces and all the planes with which we are accustomed to tame the wild profusion of existing things, and continuing long afterwards to disturb and threaten with collapse our age-old distinction between the Same and the Other.[29]

The passage simply describes a fictional classification of animals and yet it shatters Foucault's thought. On the one hand, it is a moment of clarity in which knowledge becomes limited to culture, place, and time. On the other hand, it opens upon the instability of difference. The effect extends beyond the moment, continuing to threaten the established order of knowledge. As a limit-experience, it both clarifies Foucault's knowledge and makes it tremble. "The thing that, by means of the fable, is demonstrated as the exotic charm of another system of thought, is the limitation of our own, the stark impossibility of thinking *that*."[30] Foucault felt a connection between the immediacy of individual experience and the stable background of the systems of thought that structure cultures across centuries. From this insight, it became possible to see both seemingly immobile solidity and the transformations occurring on different scales that open new possibilities.

At the heart of this realization is an aspect of order that tends to go unrecognized. On one side, there are "codes of culture" that structure language, perception, exchanges, practices, and values and on the other, philosophical and scientific theories of order that validate those codes. Between these two sides, Borges reveals that "there exists, below the level of [a culture's] spontaneous orders, things that are in fact themselves capable of being ordered . . . the fact, in short, that order *exists*."[31] To become aware that order exists, rather than being synonymous with existence itself, is to open the question of its contingency. While at a philosophical or scientific level the legitimacy of a given order can be criticized, the Borges experience reveals that order itself is a chance construction.

Order is both foundational and contingent, calling for a different mode of criticism. "This middle region, then, in so far as it makes manifest the modes of being of order, can be posited as the most fundamental of all: anterior to words, perceptions, and gestures, which are then taken to be more or less exact, more or less happy, expressions of it (which is why this experience of order in its pure primary state always plays a critical role)."[32] The experience of order in its pure state is not the experience of a real order underlying everything, but the experience of the limits and thus the reality of a specific order. Gesture and taste as well as the philosophical theories entwined with them appear as partial expressions of that order. This is how Foucault is able to show that practices of wage labor and Smith's and Marx's interpretations of those practices are based on complementary assumptions according to which labor creates value. While Smith justifies and Marx criticizes these practices, Foucault tries to push us to a point at which we might begin to see how those complementary experiences are constructed and what other possibilities might emerge.

In this way, Foucault develops genealogy as a form of "'evential' *engagement*," examining how events are connected across different scales to show what limits and possibilities an event introduces into a mode of living.[33] In doing so, Foucault substitutes "for an undifferentiated reference to *change*—which is both a general container for all events and the abstract principle of their succession—the analysis of transformations that constitute 'change.'"[34] The general notion of change reduces all events to the same form and places them within a limited temporal frame of successive occurrences. Absent a unitary notion of change evenly distributed along a chronological history, the event can no longer be seen as a self-sufficient, temporally confined break. Instead, transformations occur at different levels and temporalities. In short, history is freed from a uniform chronology of linear succession and the event is freed from an absolute image of discontinuity.

By proposing different levels of events that bring about various degrees of transformation, Foucault shows that events connect different temporal speeds and spatial scales: these overlap and interact rather than simply expressing a flat interruption in chronological succession. While Marx was able to push a theory that broke out of the frame of wealth accumulation as a primary and legitimate goal, he still relied on the labor theory of value, which continued to have explanatory power. Some of the events that Foucault explores occur on a larger timescale than the events defined by Marx's innovations. Foucault's point is that one can best understand the change that events bring about by looking at the transformations that connect those events. Even large epistemic events do not change everything since continuities subsist within them. It is a matter of discerning how the continuous and the discontinuous remain in tension with each other.

This view of how transformation occurs entails a different notion of individual agency.

> [A] change in the order of discourse does not presuppose 'new ideas', a little invention and creativity, a different mentality, but transformations in a practice, perhaps also in neighbouring practices, and in their common articulation. I have not denied—far from it—the possibility of changing discourse: I have deprived the sovereignty of the subject of the exclusive and instantaneous right to do so.[35]

The notion of the sovereign individual who makes things happen in accord with its will, central to the liberal justice theories discussed in Chapter 1, no longer holds up. Instead, the subject emerges as a site at which temporally dispersed knowledges and practices meet. Much of what the subject thinks, says, and does is a matter of contending and overlapping cultural codes of truth and regimes of action being exercised through that particular node. If the subject does not exclusively determine itself, particularly with regard to transformations, it is because events also express their effects through subjects and provide opportunities for the subject to help shape transformations in itself and society. This gives a fuller picture of the way Foucault envisioned the potential of limit-experiences as they come into contact through the subject with the deeper social and cultural milieu.

For Foucault, this is not just a methodological question but also a political one. The refusal of approaches that challenge progressive-teleological history and the unified subject are political refusals.[36] Against this refusal, he adopts an approach that emphasizes the political weight that discourses, knowledge, and practices have in our lives, as well as the events that enable possible transformations to different ways of being. He suggests some of those possibilities by linking the micro-event of his Borges limit-experience with the macro-events that dissolved and produced a series of western *epistemes*, including the contemporary one.[37] Just as Foucault's mode of being was interrupted by reading Borges, so have the *epistemes* structuring perception and understanding been interrupted by great events. But the two are connected: the micro-event of a limit experience helps to perceive epistemic macro-events. Foucault does not just "explain" the transition between different *epistemes* but brings us to the point where we can experience the limits of the contemporary Western *episteme* and the mobilization of another. This contemporary *episteme* is defined by the event of man.[38]

In tracing the outlines of the modern event of man, Foucault also outlines the fissures that might carry us beyond it. Foucault sums up this critical movement:

> From the limit-experience . . . to the order of things . . . It was upon this threshold [of modernity] that the strange figure of knowledge called man first appeared and revealed a space proper to the human science. In attempting to uncover the deepest strata of Western culture, I am restoring to our silent and apparently immobile soil its rifts, its instability, its flaws; and it is the same ground that is once more stirring under our feet.[39]

Foucault draws on the critical project of transformation that runs throughout his work to destabilize the modern *episteme*. Its apparently solid foundation excludes the kind of experiences that Foucault finds important for understanding different ways of being. The modern promise of an expanding universal knowledge that has no need of difference is the target of Foucault's critique.

If the modern *episteme* is defined by the invention of man, it is because humans are both an object of positive knowledge and that in relation to which knowledge of all other things is constituted. It is thus not just man that defines modern knowledge, but "man and his doubles."[40] Modern knowledge is structured around a reflexive twist according to which man is "a being whose nature . . . is to know nature."[41] Man is part of the natural order and his role within it is to understand that order. Yet this capacity that gives humans substance is also a limit. Foucault calls this condition that of the "enslaved sovereign"[42] because even though he is the knower of nature, it is only the texture of his experience, his own nature, that he explores. Empirical investigations point back toward the transcendental capacities of the knowing subject. Humans are trapped in this system insofar as they only "challenge" the limits of the modern order by expanding them through producing "new" knowledge. Though man experiences this inquiry as freedom, he is trapped between knowledge and experience in such a way that discoveries on one side already correspond to what was presumed to be possible on the other side. What is avoided in this back and forth is the very fact of limited human nature.

Foucault's concern is that the oscillation between these two poles has rocked thought into an "anthropological sleep . . . a sleep—so deep that thought experiences it paradoxically as vigilance, so wholly does it confuse the circularity of a dogmatism folded over upon itself in order to find a basis for itself within itself with the agility and anxiety of a radically philosophical thought."[43] Meticulous empirical observation and rigorous transcendental formalization may seem independent and even opposed projects. Yet what Foucault shows is that they move in a co-dependent circle around the figure of the human in which neither uncovers something that challenges that figure. It is not just a matter of the discourse deployed in this *episteme*, but of its "pathos" as well. The intensity, care, and rigor with which knowledge is pursued also protects the human from knowledge that might unsettle it, shielding it from the disruption of the event.

Yet if the birth of man is the event that structures the modern *episteme*, this also means that "man as such is exposed to the event."[44] If modern knowledge and existence are constituted primarily through the human, then events that interrupt and modify modern knowledge and practices also erode and transform the human. For Foucault, whether such events are warded off or engaged is a matter of how one approaches finitude. "At the foundation of all the empirical positivities, and of everything that can indicate itself as a concrete limitation of man's existence, we discover . . . a fundamental finitude which rests on nothing but its own existence as fact."[45] Finitude determines both the boundaries of positive knowledge as well as the bare existence of the being constituted as human. On the one hand, that which is beyond the limit continues to lure human activity. On the other hand, a more critical approach engages the concrete limitation of human finitude rather than viewing finitude as an as-yet-unreached horizon.

To begin to wake from the anthropological sleep, Foucault asks a different and "aberrant" question. "Does man really exist? To imagine, for an instant, what the world and thought and truth might be if man did not exist, is considered to be merely indulging in paradox."[46] Just as for Kierkegaard, the event that produces faith is aberrant and paradoxical to understanding, so for Foucault the event that challenges the given epistemic model is aberrant and paradoxical. Because of how deeply rooted the human is, breaking away from it requires an imaginative, and seemingly illogical project of questioning. To engage the event that will erode the human, Foucault builds on his reading of Nietzsche in "A Preface to Transgression." He argues that Nietzsche's thought draws on finitude in a way that challenges the modern *episteme* "when it introduces in the form of an imminent event, the Promise-Threat, the notion that man would soon be no more."[47] The promise is that the end of the modern way of understanding ourselves and the world will come and is already underway. The threat is that this entails the end of the human as we know it. The events explored in Chapter 2 attest to this: the seasonal interruption and direction of life, the affective intensity of the Return that shatters the subject, and the event of the human as a minor event for nature.

Foucault concludes *The Order of Things* by pointing to a few characteristics in three counter-sciences that had begun a transformation away from the anthropological sleep by attending to the event: psychoanalysis, ethnology, and linguistics/literature. In fact, Foucault seems to see *The Order of Things* as a critical combination of the three in a form of linguistics that draws upon the most important aspects of psychoanalysis and ethnology. First, the counter-sciences operate upon "a perpetual principle of dissatisfaction, of calling into question" what seems to be established.[48] Second, they focus on the limit as a condition of possibility for man's mode of being.[49] This is a positive orientation toward the event as a break that expresses human and epistemic finitude. Third, they attempt to produce the clarity/impossibility of limit-experiences to "dissolve man," pushing subjects to lose their consistency.[50] Fourth, the counter-sciences all incorporate an element of practice.[51] While these elements found their way into Foucault's later work, an unexpected event ended up pushing him to elaborate an ethos of the event that was simultaneously more concrete and more spiritual.

Before turning to that, however, we might speculate about other ways out of the modern *episteme*. Foucault ended *The Order of Things* with a surprisingly prescient prophecy: "if some event of which we can at the moment do no more than sense the possibility—without knowing either what its form will be or what it promises—were to cause [the current arrangements of knowledge] to crumble . . . then one can certainly wager that man would be erased, like a face drawn in sand at the edge of the sea." (422) Since it was not possible to know the form of this event, let us ask whether it might be the Anthropocene. After all, if one were only "sensing" the limits of the modern project, the environment would certainly be one source of tremors. Initially, the notion of the Anthropocene seems to go in exactly the opposite direction, inscribing the human even more profoundly at the heart of nature and knowledge. It elevates the planet-altering power of humanity while intensifying a research program to complete the knowledge of the Earth system to find out how this is possible and what can be done. But a variety of new approaches including new materialist philosophies, complexity theory, and those attending to nonhuman agency seem to challenge the anthropological sleep and may even be nascent expressions of a different *episteme* altogether. These approaches have been critical for thinkers like Donna Haraway, William E. Connolly, and Bruno Latour in their engagements with the Anthropocene event. In accord with his suggestion that change does not follow a set model, new forms of knowledge and practices unexpected by Foucault may be reworking the anthropological modern *episteme*.

Spiritual resolve: the politics of truth as a way of life

Foucault's intellectual trajectory was interrupted by an unexpected event in the form of the Iranian revolution, which emerged "like a *thunderbolt from the blue*."[52] This event challenged existing institutions and ways of understanding the world, including that of Foucault, whose thought was transformed through his attempt to make sense of it. Despite his lack of expertise in Iranian history and Islamic theology, a number of aspects of his thought, such as his commitment to non-teleological history and his focus on the possibility of novel transformation through the event, enabled him to recognize the liberatory potential of the revolution outside of a Western, Enlightenment framing. Even more, as Behrooz Ghamari-Tabrizi shows, what Foucault saw in Iran drove the shift to questions of political spirituality in his late thought.

Though Foucault was influenced in a number of ways by his experience giving a journalistic account of the Iranian revolution, Ghamari-Tabrizi emphasizes the significance of Foucault's discovery of the collective will. As Foucault remarked in an interview:

> Among the things that characterize this revolutionary event, there is the fact that it has brought out—and few people in history have had this—an absolutely collective will. The collective will is a political myth . . . and, personally, I thought that the collective will was like God, like the soul, something one would never encounter. I don't know whether you agree with me, but we met in Tehran and throughout Iran, the collective will of a people.[53]

Ghamari-Tabrizi calls this experience a "transformative moment."[54] As I have argued, the transformative aspect of the event—whether it takes the form of a limit experience or larger break—was always the most important one for Foucault. The transformation that Ghamari-Tabrizi effectively traces is one in which Foucault was pushed to reconsider the ethics of political subjectivity through witnessing individual and collective revolutionary becoming in Iran.

Many of the key aspects of Foucault's later work have formative elements in this experience, such as the courage of people on the streets even as they were attacked and the connection between spirituality, truth, and emancipation.[55] Perhaps the most important revolutionary characteristic Foucault saw in Iran, Ghamari-Tabrizi argues, was "the possibility of resistance without participating in or perpetuating a *preconceived* schema of power."[56] Thus, Foucault focused on revolutionary becoming as a way to open history to new possibilities rather than an expected outcome. This is why he was able to appreciate the possibility of the Iranian revolution without feeling compelled to criticize the entire movement on the basis of its outcome. This "critical ontology of the present" examines "the way we engage and experience our life circumstances . . . to recognize and promote a spiritual and ethical self who is willing to pay the price of a transformative engagement with his or her actuality."[57] Political resistance is not a matter of laying out a plan to be implemented but of developing a self that can effectively connect to and transform the present. Thus, Ghamari-Tabrizi concludes that Foucault's return to Kant is not a return to universal Enlightenment humanism, but an escape from any universal or predetermined notion of the human through the irreducibility of the subject who revolts.

In this section, I will take up this notion of a resistant subject by examining how in Foucault's return to the ancients he developed an ethos of the event as a political technique. The object will be to see how such an ethos might invigorate a politics of responsiveness to climate change. Yet the point is not to take revolutionary Iran, the ancients, or even Foucault himself, as an example to be replicated. As Foucault argued, "you can't find the solution of a problem in the solution of another problem raised at another moment by other people."[58] Rather, the point is to think about how the event must be worked into the constitution of the self as a node of thought and action. For Foucault, it was no longer enough to either work through the sporadic intensity of the limit-experience or develop a genealogical view of transformation. Through the care of the self, he sought to develop an ethos of the event that would have the effect of making the subject more worldly, aligning its beliefs and actions, and serving as a form of political engagement.

Caring for the self is an ethics that problematizes the liberal image of freedom. Freedom is not an innate right or capacity of every individual but rather can only be actualized through ethos.

> *Ēthos* . . . was a mode of being for the subject, along with a certain way of acting, a way visible to others. A person's *ēthos* was evident in his clothing, appearance, gait, in the calm with which he responded to every event, and so on . . . But extensive work by the self on the self is required for this practice of freedom to take shape in an *ēthos* that is good, beautiful, honorable, estimable, memorable, and exemplary.[59]

For Foucault, the subject is composed of a number of different elements including practices, impulses, habits, principles, and aspirations. Further, this subject is multiple, being "the subject of instrumental action, of relationships with other people, of behavior and attitudes in general, and the subject also of relationships to oneself."[60] Caring for the self is a matter of developing an ethos to shape and maintain this complex interplay in the self. The goal is to see to what degree the subject can free itself by forging a consistency that extends from walking to larger events that interrupt life. As I showed in Chapter 1, Kierkegaard also thinks that someone who experiences an event and responds to it with faith expresses a particular appearance and gait that are linked to their underlying mode of being. While for Kierkegaard, faith produces a singular mode of being, for Foucault, this will always be incomplete and continue to change.

What interests Foucault in this notion of ethos is that it is a way of transforming the self and even of making transformation a constant task for the self. This is a continuation of the task of becoming other that Foucault develops in "A Preface to Transgression" and *The Order of Things*. Folding the concept of caring for oneself into the task of becoming other gives it a new articulation in terms of "spiritual truth." In spiritual truth, "for the subject to have right of access to the truth he must be changed, transformed, shifted, and become, to some extent and up to a certain point, other than himself. The truth is only given to the subject at a price that brings the subject's being into play."[61] This is a more robust theory of subjective transformation but also more difficult and riskier. It entails the removal of characteristics and modes of action that define who one is such that one cannot simply abandon the new self and return to a previous state.

The transformation can be produced through exercises (*askēsis*) that equip (*paraskeuē*) the self, enabling the subject to achieve the truth. These exercises help transform reasoned principles of truth into the actions undertaken by the subject.

> The *paraskeuē* is, again, the element of transformation of *logos* into *ethos*. And the *askēsis* may then be defined as the set, the regular, calculated succession of procedures that are able to form, definitively fix, periodically reactivate and, if necessary, reinforce this *paraskeuē* for an individual. The *askēsis* is what enables truth-telling—truth-telling addressed to the subject and also truth-telling that the subject addresses to himself—to be constituted as the subject's way of being.[62]

Caring for oneself transforms truths reached by reason into an ethos. Thus, principles become ingrained in character through a spiritual transformation. A new mode of being is produced in which truth-telling is not a matter of conveying knowledge but of expressing a conviction through a set of actions. Truth is not final, but that which is true for a particular subject at a given time open to yet other transformations and truths through further practice and experimentation.

It is worth more closely examining the practice of *paraskeuē* to understand Foucault's ethos of the event. *Paraskeuē* is a set of practices that actively equip and prepare the self to better encounter unforeseen events. The emphasis on active training is made through a comparison with the athlete. "The Stoic athlete, the athlete of ancient spirituality also has to struggle. He has to be ready for a struggle in which his adversary is anything coming to him from the external world: the event. The ancient athlete is an athlete of the event."[63] The practice involves hearing, understanding, learning, repeating, and memorizing a series of discourses (*logoi*). These discourses are ideas, truths, phrases, behavioral guides, and principles. "Discourses should be understood as statements with a material existence" that become a "permanent virtual and effective presence, which enables immediate recourse to them when necessary."[64] Through learning, understanding, and memorizing these principles they become incorporated into the self. They are not true because they are known, but become true when they infuse and motivate bodies, thoughts, practices, and modes of living.

> All the verbal repetitions must be part of the preparation so that the saying can be integrated into the individual and control his action, becoming part, as it were, of his muscles and nerves . . . when the event occurs, the *logos* at that point must have become itself the subject of action, the subject of action must himself have become at that point *logos* and, without having to sing the phrase anew, without even having to utter it, acts as he ought to act.[65]

The learned and practiced discourses are not the self, but are always virtually present, ready to guide the action of the subject at crucial moments. They are preparations made for an unexpected event, with a material existence that extends into behavior. In this practice, bodily responsiveness manifests a particular attitude toward encountering the event.

This ethos of the event has a number of effects on the subject. First, it connects the subject to the world. Drawing on Seneca, Foucault highlights two sides to this process of connection. The first is a "punctualizing" technique that establishes "the maximum tension between the self as reason and the self as point."[66] This understanding is a "stepping back from the point we occupy" to simultaneously see our place and our interconnection with the larger world where each thing has its place and "to which we belong."[67] Liberated from a narrow focus on the self-sufficient individual, a sense of belonging to the larger world emerges, enabling us to "dismiss and exclude all the false values and . . . to take the measure of our existence."[68] "Second, the knowledge of nature is liberating inasmuch as it allows us . . . to ensure a *contemplatio sui* in which the object of contemplation is ourselves in the world, ourselves inasmuch as our existence is linked to a set of determinations and necessities whose rationality we understand."[69] As with limit-experiences, Foucault seeks a "maximum tension" which pushes the subject to become otherwise. This tension is liberating because it shifts our position from being unthoughtfully subject to these external forces to understanding and participating in them to some degree.

The uncertain, dynamic, and interconnected view of the world with all of its beauty, pleasure, suffering, deception, and stupidity that emerges from this requires and supports an affirmative existential disposition. "[One] is shown the world precisely so that he clearly understands that there is no choice, that nothing can be chosen without choosing the rest, that there is only one possible world, and that we are bound to this world . . . the only point of choice is this . . . Consider whether you want to enter or leave, that is to say, whether or not you want to live."[70] Seneca, from whom Foucault takes this passage, and Foucault himself both reply in the affirmative. It is notable that Kierkegaard and Nietzsche also emphasized an affirmation of belonging to this unruly world with its events, tragedies, joys, and uncertainty. Schmitt is the only thinker examined in this study who perhaps was not able to affirm such a world. Though he enjoyed many aspects of existence and did not shy away from acknowledging its problematic parts, he likely saw the world as too precarious to feel at home in it. Thus, his thought was driven away from the riskier aspects of life and toward false promises of security and consolation. An affirmative ethos of the event involves an existential commitment to belonging to this world in its entirety.

The second effect is that caring for oneself allows this ethos to manifest as a way of living. Practices and principles reach into the regularized habits and expectations of daily life. "The *tekhnē tou biou*, the way of dealing with the events of life, must be inserted within a care of the self that has now become general and absolute. . . . One lives with the relationship to one's self as the fundamental project of existence, the ontological support which must justify, found, and command all the techniques of existence."[71] Daily living becomes a project. Diet, occupation, media choices, aspirations, travel, the orientation to death, housing, interpersonal relations, etc. all become important in caring for oneself. Affirming a world to which reason is insufficient entails a supplementary set of practices and principles that become an existential foundation. Preparing for and responding to events require continuously attending to and trying to shape the motions of life.

Third, this establishment of care of the self as a way of living in an unruly world expresses the truth obtained through spiritual transformation. Foucault argues that for the ancients it was a matter of "knowing the extent to which the fact of knowing the truth, of speaking the truth, and of practicing and exercising the truth enables the subject not only to act as he ought, but also to be as he ought to be and wishes to be."[72] The result of caring for oneself is that one experiences and lives the truth that they know. Foucault uses *parrhēsia* to elucidate this. *Parrhēsia* is a form of particularly free and open speech that is characterized primarily by a realization of spiritual truth. "What must be shown is not just that this is right, the truth, but also that I who am speaking am the person who judges these thoughts to be really true and I am also the person for whom they are true . . . What characterizes *parrhēsia*, *libertas*, is this perfect fit between the subject who speaks, or the subject of enunciation, and the subject of conduct."[73] A conduct or mode of being becomes a living truth because it has been developed and incorporated by a subject who speaks on the truth of it. That person vouches for those truths with their existence since it really is the life that they are living. Contrarily, when one's actions and their claims about truth do not line up, it is clear that one is not caring for oneself. Such a test serves a clear and critical function with regard to those who say they believe in climate destabilization but do not live accordingly. This has a particularly critical function today, when even though many people know about climate change, they do nothing about it.

Finally, care of the self is political. Foucault discusses the political implications of caring for oneself in the ancient world a number of times, but he also argues that it is an essential political task today. "I think we may have to suspect that we find it impossible today to constitute an ethic of the self, even though it may be an urgent, fundamental, and politically indispensable task, if it is true after all that there is no first or final point of resistance to political power other than in the relationship one has to oneself."[74] Foucault suggests that the various versions of contemporary self-development—being oneself, freeing oneself, etc.—are all blocked and ossified efforts. This is in part due to the forms of governmentality under which we exist. Indeed, it seems that there are fruitful connections between Foucault's development of care of the self and his critique of the way neoliberalism pushes people to make themselves into entrepreneurs of the self.[75] But for Foucault, governmentality always has to pass through the relation one has with oneself.

This is why Foucault thinks the self is one important node of political intervention. If caring for oneself is "urgent, fundamental, and politically indispensable," it is because the contemporary political milieu deploys individual development and success as a way to further systematic inequality and exclusion, and we have few counter-notions of the self with which to resist it. Amidst the network of cultural and social imperatives and incentives to be individually economically successful, to consume any of a myriad of goods and lifestyles, to draw sharp lines between us and those who threaten our safety and well-being, to promote certain forms of strength and power, to adhere to a calculable short-range view of time, and so on, the relationship that one has to oneself can provide a checkpoint for these comforts, values, and impulses. Caring for oneself, insofar as it brings the spiritual dimension into play, insofar as it involves risking one's mode of being, is already a political mode of engagement.

Why risk yourself when so many comforts and goods are available? As Foucault indicates, contemporary modes of living militate against spiritual truths, against considering and recognizing events that interrupt and upset our modes of being, and against occupying oneself with techniques and exercises. Caring for oneself can thus be seen as what he called an "ethic of discomfort."

> Never to consent to being completely comfortable with one's own presuppositions. Never to let them fall peacefully asleep, but also never to believe that a new fact will suffice to overturn them; never to imagine that one can change them like arbitrary axioms, remembering that in order to give them the necessary mobility one must have a distant view, but also look at what is nearby and all around oneself.[76]

Foucault argues for a self-practice that critically engages presuppositions in order to transfigure the self, since facts are insufficient for that. Indeed, the biggest criticism of the climate skeptics all too often seems to be that they are unable to accept a new fact. But at the same time, that fact has proven insufficient for changing the lives of many of those who do accept it. It is neither convenient nor comfortable to have to adapt the future you imagined for yourself, your community, and the world to a new circumstance, particularly when the presuppositions about the world that enable that future militate against that change. Yet such transformations become possible through an ethos that connects a dynamic subject to an uncertain world.

Responding to climate change through changes in personal behavior has rightly been criticized. Amitav Ghosh gives a particularly compelling critique when he argues that it gets caught in a politics of sincerity in which the credibility of broader calls for government action founder amidst inquisitions into individual lifestyle and personal sacrifice.[77] But the problems of a politics of personal purity are not limited to those who seek to cleanse themselves of their climate contribution. William T. Vollmann uses his reliance on fossil fuels and lack of responsiveness to climate change to validate his analysis of how deeply embedded society is in carbon-based energy.[78] His purity consists in not being the kind of hypocrite who demonstrates how much society relies on fossil fuels while having himself reduced his own usage. Yet this lack of hypocrisy is the other side of the existential complicity it enacts with the carbon machine. The ethics of the subject put forward by Foucault avoids both of these traps of purity. On one hand, because the subject is never complete and never master of itself, it is never a question of a final way of living that yields an authentic self. Rather, how to live is always an active task. On the other, one can never be comfortable with an existing way of being. Embracing the flaws of a given political arrangement as a kind of truth of the fallen can never be a source of assurance. Finally, for Foucault, the individual is not the unit of analysis but the individual in relation to a socio-political reality. If the subject has to take itself as an object and a task, it is to resist and change those institutions and ways of ordering society that have produced it.

***

In resonance but not direct connection with Foucault, Paul Kingsnorth is developing an ethos that may be exemplary for responding to climate change. After 20 years of environmental activism, he withdrew from the movement to try to understand and articulate the environmental crisis we face. With Dougauld Hine, Kingsnorth laid the groundwork for this attempt in "Uncivilization: The Dark Mountain Manifesto." This document founded the Dark Mountain Project as a way to sound out the climate event and the place of humanity amidst increasing environmental destruction.

The Dark Mountain project is about being able to tell the truth. It begins with the attempt to get away from self-betrayal. "I do feel the need to be honest with myself, which is where the 'walking away' comes in. I am trying to walk away from dishonesty, my own included."[79] Walking away makes it possible to audit our situation and find a way to articulate what is happening. "And so we find ourselves, all of us together, poised trembling on the edge of a change so massive that we have no way of gauging it. None of us knows where to look, but all of us know not to look down . . . . We believe it is time to look down."[80] Among those who refuse to look down, are politicians, business people, those who continue shopping, those who simply despair, and environmentalists who "work frantically to try and fend off the coming storm."[81] Looking down is not about facing the facts and stating them or giving a forced solution. Rather, the main project is to tell stories in light of a narrative failure to make sense of our situation, "to examine this process, and our place in it, and to do so from beyond the framework of our current cultural assumptions."[82] Just as Foucault suggests, the need to create subjects capable of telling the truth, Kingsnorth relocates himself politically, culturally, and existentially to be able to tell a story that confronts our contemporary predicament.

As a crucial part of the project, Kingsnorth and Hine attempt to "stand outside the human" with what they call "uncivilised art." "It sets out to paint a picture of homo sapiens which . . . [another] being from our own [world] . . . might recognise as something approaching a truth."[83] They seek to produce and curate art, and writing in particular, that turns away from the myths and stories that have propelled western humanity for the last few centuries to locate the place of the human within nature. Yet this truth can only be stated through "the shifting of emphasis from man to notman."[84] We might understand this as an attempt to express a fundamental knowledge about human limitation and finitude outside of the endless knowledge accumulation that constitutes the anthropological sleep.

This position is underscored in a comment that Kingsnorth made in response to Wen Stephenson, who argued that a better course of action than withdrawing would be that of Tim DeChristopher who went to prison for committing fraud to outbid oil companies for federal land use rights. DeChristopher justified his actions by saying "I would never go to jail to protect animals or plants or wilderness. For me, it's about the people."[85] Wen sees this as "a humanitarian imperative [that] transcends environmentalism and environmental politics."[86] Kingsnorth's response contests this imperative. "The Tim DeChristopher quote which you use approvingly is something which divides us . . . I'm of the opinion that the last thing the world needs right now is more "humanitarians." What the world needs right now is human beings who are able to see outside the human bubble, and understand that all this talk about collapse, decline, and crisis is not just a human concern."[87] Kingsnorth is trying to undercut the assumptions that privilege the human, which is part of what makes his politics so frightening and offensive to people. The idea is that we cannot find out what place in nature we really occupy until we give up on the assumption that humans are at the center of it.

Following in the tradition of caring for oneself, Kingsnorth adopts a few experimental exercises to help him carry out this task. One is to

> take part in a very ancient practical and spiritual tradition: withdrawing from the fray . . . Withdraw so that you can allow yourself to sit back quietly and feel, intuit, work out what is right for you and what nature might need from you. Withdraw because refusing to help the machine advance—refusing to tighten the ratchet further—is a deeply moral position. Withdraw because action is not always more effective than inaction. Withdraw to examine your worldview: the cosmology, the paradigm, the assumptions, the direction of travel. All real change starts with withdrawal.[88]

When he withdrew, he was heavily criticized as a nihilist, a defeatist, and a romantic, among other things. But Kingsnorth frames withdrawing as an alternative form of action. It is a refusal to participate in a destructive culture. Being in the impossible position of wanting to avert climate change and yet still embodying climate change by being a living member of a particular ecological milieu pushed

Kingsnorth to a different position. He had to withdraw to understand this position and what to do about it. Clearing some space for this process is a critical source of further change.

Another exercise that Kingsnorth practices and teaches is scything. This activity produces a different mindset and opens a different level of connection with nature.

> Using a scythe properly is a meditation: your body in tune with the tool, your tool in tune with the land. You concentrate without thinking, you follow the lay of the ground with the face of your blade, you are aware of the keenness of its edge, you can hear the birds, see things moving through the grass ahead of you. Everything is connected to everything else, and if it isn't, it doesn't work. Your blade tip jams into the ground, you blunt the edge on a molehill you didn't notice, you pull a muscle in your back, you slice your finger as you're honing. Focus—relaxed focus—is the key to mowing well.[89]

When writing about this exercise, Kingsnorth intersperses it with reflections on the writings of Ted Kaczynski, who he both admires and resists. But he continues on to further reflections on the failure of the environmental movement, its resonance with neoliberalism, technology fetishism, convivial modes of living, time, progress, and a possible course of action. He ties all of these to the activity of scything. Much like some of the exercises that Foucault cites, this is on one hand a repetitious training of the body and mind, and on the other hand a way of producing new thinking that alters behavior.

These exercises are part of a spiritual transformation. Kingsnorth resists organized religion and new-age notions of the sacred. Yet if "[t]he Dark Mountain Project arose out of a collapse in belief,"[90] then what Kingsnorth noticed afterward was the importance of spiritual experiences in nature. For him, spending time in the wilderness is a key aspect of undergoing spiritual transformations through which humans come to understand their place in and connection to the rest of the world. Further, Kingsnorth calls for those who feel that nature is sacred to try to speak truthfully about it. "This feeling is not an awkward and embarrassing stumbling block in the way of a rational assessment of the reality of ecosystems . . . And those of us who do feel it . . . have a duty to talk about it, openly, calmly, incisively . . . we should at least try and find the words for what is so plainly missing. This is not an indulgence, but a necessity."[91] In the mode of *parrhēsia*, Kingsnorth argues that it is a *necessity* to speak a spiritual truth of connection to nature that doesn't conform to reason, even though the cultural milieu is dismissive toward this perspective.

Kingsnorth's approach molds practices and the transformations they produce into a way of life. He conceives of climate change as an existential problem. "We are all climate change. It is not the evil "1%" destroying the planet. We are all of us part of that destruction. This is the great, conflicted, complex situation we find ourselves in. I am climate change. You are climate change. Our culture is climate change. And climate change itself is just the tip of a much bigger iceberg."[92]

Indeed, for him this denial "extends to every aspect of what we produce and how we live our lives."[93] Taking seriously the problem that everyone, particularly in the West, drives climate change and environmental destruction through their life activities means that a different way of living has to be produced. The problem is an existential failure "to distinguish between life and lifestyle."[94] The Dark Mountain Project seeks to bring out this distinction and to give new meaning to contemporary existence. They hope to "redraw the maps . . . by which we navigate all areas of life . . . Our maps must be the kind sketched in the dust with a stick, washed away by the next rain."[95] If the existing stories only propel us along a fateful and false course, then new ones are needed which reshape life. Such stories are not limited to specific political and scientific solutions or a set course of action, but should inspire developments and experiments that try to connect humans to nature. Like Foucault's notion of a subject that periodically undergoes change, these maps are temporary and should install new behaviors while also allowing for future changes as well. In line with this, Kingsnorth recently left the Dark Mountain Project after an ascetic venture into the woods for four days with no food and minimal shelter. He realized that it was an important path to take, both to let the project keep transforming and to let himself take on other experiments.

Finally, this project entails a new connection to the world. Kingsnorth wants to accept and love the world, with all the good and bad that it entails. Setting out on this path means accepting and affirming the worst possible outcomes of climate change, even as one might fear them and want to work to prevent them. "The Uncivilised writer knows the world is . . . something we are enmeshed in—a patchwork and a framework of places, experiences, sights, smells, sounds."[96] Even if we live amidst these sounds and smells, we do not experience them and give them weight and meaning. The task for Kingsnorth and others like him is to make that world tangible and thinkable, to find ways to connect themselves and to reach out to help others connect themselves as well. Reforging this connection means giving locality contour, depth, and meaning, and forging a vital and spiritual link with nature as it sustains and perhaps destroys us. But it is also a matter of forging new social connections to nature, as Kingsnorth's work demonstrates. Drawing on Foucault, William E. Connolly develops the relevance of the specific intellectual, "whose technical skills and specific capacities form a niche that have become strategic during this era."[97] Kingsnorth, like Naomi Klein and James Hansen and Foucault too, is a specific intellectual. Such intellectuals then become part of a broader "politics of swarming" which is not a unified movement, but whose different actors and segments have "some potential to augment and intensify the others."[98] This is one image of what a Foucauldian politics of caring for oneself might look like today.

Kingsnorth's path is not the only one. But we might at least start with the broad orientation that Foucault uses. "I am fascinated by history and the relationship between personal experience and those events of which we are a part."[99] From limit-experiences to caring for oneself, we need ways to engage the event. Because their effects can be so dramatic, because they are hard if not impossible to control, and because they can arrive unexpected or go unobserved, events play an uncertain and volatile role. This danger is central to Foucault's approach. "My point is not that everything is bad, but that everything is dangerous, which is not exactly the same as bad. If everything is dangerous, then we always have something to do. So my position leads not to apathy but to a hyper- and pessimistic activism. I think that the ethico-political choice we have to make every day is to determine which is the main danger."[100] What is the most pressing, dangerous event today? Sounding it out indicates some of the dangers we face and the destructive human-centeredness of how we are living. Yet this new alertness is only a starting point for diagnosing our relations and the various dangers we confront today. Responding to this event will require alertness, combined with a willingness to endure discomfort, sacrifice, and transformation, and the existential energy to experiment until we are able to live more appropriately in relation to the problems we face.

Notes

1 Sloterdijk, *Philosophical Temperaments*, 98–9.
2 Oreskes and Conway, *The Collapse of Western Civilization*, 1–2, 35–6.
3 Ibid., 48–9.
4 Latour, "On Some of the Affects of Capitalism."
5 Foucault, "Questions of Method," 76–8.
6 Ibid.
7 Ibid.
8 Ibid., 82.
9 Foucault, "Interview with Michel Foucault," 241–2.
10 Ibid., 241.
11 Ibid., 244.
12 Ibid., 242.
13 Ibid., 240.
14 Ibid., 244–5.
15 Ibid., 242.
16 Ibid., 243.
17 Ibid., 245–6.
18 Jay, "The Limits of Limit-Experience," 160–1.
19 Ibid., 159.
20 Ibid., 168.
21 Foucault, "A Preface to Transgression," 85.
22 Ibid., 71.
23 Nietzsche, *The Gay Science*, 119–20.
24 Foucault, "A Preface to Transgression," 72.
25 Ibid., 73.
26 Ibid., 74.
27 Ibid., 75. In the second phrase, I have used the word "upon" from the original translation by Daniel F. Bouchard and Sherry Simon rather than the word "about," to which it was changed in the Rabinow edition.
28 Foucault, "Preface to Transgression," 75.
29 Foucault, *The Order of Things*, xv.
30 Ibid.
31 Ibid., xx.
32 Ibid., xxi.

33 Foucault, *The Archaeology of Knowledge*, 168.
34 Ibid., 172–3.
35 Ibid., 209.
36 Ibid., 209–10.
37 Foucault, *The Order of Things*, xxi–xxii.
38 Ibid., 317, 345.
39 Ibid., xxiv. My brackets.
40 Ibid., 303.
41 Ibid., 310.
42 Ibid., 312.
43 Ibid., 341.
44 Ibid., 370.
45 Ibid., 315.
46 Ibid., 322.
47 Ibid.
48 Ibid., 373.
49 Ibid., 376–8.
50 Ibid., 375, 377, 379.
51 Ibid., 376, 382.
52 Ghamari-Tabrizi, *Foucault in Iran*, 20.
53 Foucault, "Iran: The Spirit of a World Without Spirit," 253.
54 Ghamari-Tabrizi, *Foucault in Iran*, 58, 63.
55 Ibid., 56–8, 180–2.
56 Ibid., 68.
57 Ibid., 167.
58 Foucault, "On the Genealogy of Ethics," 256.
59 Foucault, "The Ethics of the Concern of Self as a Practice of Freedom," 286.
60 Foucault, *The Hermeneutics of the Subject*, 57.
61 Ibid., 15.
62 Ibid., 327.
63 Ibid., 322.
64 Ibid., 322–4.
65 Ibid., 326.
66 Ibid., 278–9.
67 Ibid., 276.
68 Ibid., 277.
69 Ibid., 279.
70 Ibid., 284–5.
71 Ibid., 448.
72 Ibid., 318.
73 Ibid., 405–6.
74 Ibid., 252.
75 Foucault, *The Birth of Biopolitics*.
76 Foucault, "For an Ethic of Discomfort," 448.
77 Ghosh, *The Great Derangement*, chapter 3.
78 Vollmann, *Carbon Ideologies*.
79 Stephenson, "I Withdraw."
80 Kingsnorth and Hine, "Uncivilization."
81 Ibid.
82 Kingsnorth, "Journey to the Dark Mountain."
83 Kingsnorth and Hine, "Uncivilization."
84 Ibid., citing Robinson Jeffers.
85 Stephenson, "I Withdraw."
86 Ibid.

87 Ibid.
88 Kingsnorth, "Dark Ecology."
89 Ibid.
90 Kingsnorth, "Journey to the Dark Mountain."
91 Kingsnorth, "In the Black Chamber."
92 Stephenson, "I Withdraw."
93 Kingsnorth, "Journey to the Dark Mountain."
94 Ibid.
95 Kingsnorth and Hine, "Uncivilization."
96 Ibid.
97 Connolly, *Facing the Planetary*, 125.
98 Ibid.
99 Foucault, "An Interview by Stephen Riggins," 124.
100 Foucault, "On the Genealogy of Ethics," 256.

Bibliography

Connolly, William E. *Facing the Planetary: Entangled Humanism and the Politics of Swarming*. Durham: Duke University Press, 2017.

Foucault, Michel. *The Archaeology of Knowledge*. Translated by A. M. Sheridan Smith. New York: Vintage, 2010.

Foucault, Michel. *The Birth of Biopolitics*. Edited by Michel Senellart. Translated by Graham Burchell. New York: Palgrave Macmillan, 2008.

Foucault, Michel. "The Ethics of the Concern of Self as a Practice of Freedom." In *Ethics: Subjectivity and Truth*, edited by Paul Rabinow. Translated by P. Aranov and D. McGrawth and amended by Robert Hurley. New York: The New Press, 1997.

Foucault, Michel. "For an Ethic of Discomfort." In *Power*, edited by James Faubion. Translated by Robert Hurley. New York: The New Press, 2000.

Foucault, Michel. *The Hermeneutics of the Subject*. Edited by Frédéric Gros. Translated by Graham Burchell. New York: Picador, 2005.

Foucault, Michel. "An Interview by Stephen Riggins." *Ethics: Subjectivity and Truth*, edited by Paul Rabinow. New York: The New Press, 1997.

Foucault, Michel. "Interview with Michel Foucault." In *Power*, edited by James D. Faubion. Translated by Robert Hurley and others. New York: The New Press, 2001.

Foucault, Michel. "Iran: The Spirit of a World without Spirit." In *Foucault and the Iranian Revolution: Gender and the Seductions of Islamism*, edited by Janet Afary and Kevin B. Anderson. Chicago: University of Chicago Press, 2005.

Foucault, Michel. "On the Genealogy of Ethics." In *Ethics: Subjectivity and Truth*, edited by Paul Rabinow. Translated by Robert Hurley and others. New York: The New Press, 1997.

Foucault, Michel. *The Order of Things*. New York: Vintage, 1994.

Foucault, Michel. "A Preface to Transgression." In *Aesthetics, Method, and Epistemology*, edited by James Faubion. Translated by Donald F. Bouchard and Sherry Simon and modified by Robert Hurley. New York: The New Press, 1998.

Foucault, Michel. "Questions of Method." In *The Foucault Effect*, edited by Graham Burchell, Colin Gordon, and Peter Miller. Translated by Colin Gordon. London: Harvester Wheatsheaf, 1991.

Ghamari-Tabrizi, Behrooz. *Foucault in Iran: Islamic Revolution After the Enlightenment*. Minneapolis: University of Minnesota Press, 2016.

Ghosh, Amitav. *The Great Derangement: Climate Change and the Unthinkable*. Chicago: University of Chicago Press, 2016.

Jay, Martin. "The Limits of Limit-Experience: Bataille and Foucault." *Constellations* 2, no. 2 (1995): 155–74.

Kingsnorth, Paul. "Dark Ecology." *Orion Magazine*, January–February 2013. www.orion magazine.org/index.php/articles/article/7277.

Kingsnorth, Paul. "In the Black Chamber." http://paulkingsnorth.net/journalism/in-the-black-chamber/.

Kingsnorth, Paul. "Journey to the Dark Mountain." www.paulkingsnorth.net/journalism/journey-to-the-dark-mountain/.

Kingsnorth, Paul, and Dougald Hine. "Uncivilization: The Dark Mountain Manifesto." http://dark-mountain.net/about/manifesto/.

Latour, Bruno. "On Some of the Affects of Capitalism." In *Lecture Given at the Royal Academy*. Copenhagen, February 26, 2014. www.bruno-latour.fr/sites/default/files/136-AFFECTS-OF-K-COPENHAGUE.pdf.

Nietzsche, Friedrich. *The Gay Science*. Translated by Josefine Nauckhoff. Cambridge: Cambridge University Press, 2001.

Oreskes, Naomi, and Erik M. Conway. *The Collapse of Western Civilization*. New York: Columbia University Press, 2014.

Sloterdijk, Peter. *Philosophical Temperaments*. Translated by Creston Davis. New York: Columbia University Press, 2013.

Stephenson, Wen. "'I Withdraw' a Talk with Climate Defeatest Paul Kingsnorth." *Grist*, April 11, 2012. http://grist.org/climate-energy/i-withdraw-a-talk-with-climate-defeatist-paul-kingsnorth/.

Vollmann, William T. *Carbon Ideologies*. 2 vols. New York: Viking, 2018.

6   An ethos for the climate event

In *Don't Even Think about It*, George Marshall asks how it is possible for people to know about climate change while still being unable to accept it, even as they are aware of this inability. In doing so, he questions the paradoxical behavior of even those who are well informed about climate change. "Through asking these questions" he writes, "I have come to see climate change in an entirely new light: not as a media battle of science versus vested interests or truth versus fiction, but as the ultimate challenge to our ability to make sense of the world around us."[1] Responding to climate change is not a matter of future IPCC reports with greater levels of certainty, of better educating the public about the facts of climate change, or of overcoming religiously founded denial and the influence of the fossil fuel lobby to institute a political remedy, though all of those are important. Instead, Marshall suggests that climate change is a question of how to understand ourselves and our place in the world. I argue that this auditing and reworking of our place in the world is best accomplished through an ethos that enables us to respond to the depth and disruption of climate change as we both fuel and flee it.

This chapter draws on the philosophies presented in the last five chapters to suggest what such an ethos might look like. I will begin by exploring the potentials and limits of scientific understandings of climate change. Then I will examine political-theological approaches to analyze both sovereign political responses to climate change and what it would mean to believe in climate change. At another level, an ethical-cultural take identifies problematic cultural drives that push us deeper into the climate crisis and proposes ways to transform those drives into more responsible modes of living. Finally, I will look at a tragic reading of climate change that enables us to love and participate in an unruly world that is sometimes hostile to human existence. Though none of the thinkers examined in this study can provide a formula for responding to the climate event, collectively they intensify our knowledge and experience to produce the interruptions that may help us acclimatize ourselves to it.

The limits of science: complexity, uncertainty, caution

People often imagine climate change linearly, following a dose–response model: more carbon produces a proportionate increase in temperature and sea-level rise.

This view pairs with the assumption that whenever carbon emissions are reduced, the effects will also halt at the correlate level. In *With Speed and Violence*, Fred Pearce discusses the history and science of a particularly troubling aspect of climate change: tipping points. Tipping points are lines which, once crossed, set processes in motion which are very difficult, if not impossible, to reverse. Climate change is not a linear movement, but contains triggers and switches that can alter the state and functioning of the global climate system. "man-made global warming will very probably unleash unstoppable planetary forces. And they will not be gradual. The history of our planet's climate shows that it does not do gradual change."[2] These triggers are often hidden: climate history shows that they exist, but it is hard to say where they lie and how they work. Furthermore, once they have been set off, they are likely difficult if not impossible for humans to reverse.

Since climate change is not linear, some of the effects are not slow and steady, but rapid and severe. This difficult fact orients Pearce's approach to climate change: "The central message of this book is that while skeptics about climate change have a valid point when they say that scientists' climate predictions are far less certain that is often claimed, those skeptics are dreadfully wrong to take comfort in this. I take no comfort at all. There is chaos out there, and we should be afraid."[3] It is certain that Earth's climate system underwent rapid changes with enormous and sometimes devastating effects well before the age of the Anthropocene, when human effects on climate became prominent. However, when it comes to the state of the climate system today and the effect of human behavior on it, we are confronted with uncertainty. Pearce points out that for some, that uncertainty itself is a solution to the problem: there is no climate problem until we know exactly what it is. For Pearce, however, uncertainty intensifies the problem, accentuating the ways in which human life exists in connection, tension, and fragility with the larger world.

Pearce looks at historical, contemporary, and potential future climate events to situate humans in the world. One significant historical event is the onset of the last ice age 12,800 years ago, which was triggered by a large glacial lake in North America emptying into the Atlantic Ocean. It may be that this onslaught interrupted the regularity of the ocean conveyer system which circulates water throughout the world. That, in turn, disrupted the existing climate. "Within about a generation, temperatures fell worldwide—perhaps by as little as 3 to 5°F in the tropics, but by an average of as much as 28 degrees farther north, and . . . by 54 degrees in winter at Scoresby Sound, in eastern Greenland."[4] After 1300 years, the freeze ended "and temperatures returned to their former levels even faster than they had fallen. . .'Most of that change looks like it happened in a single year. It could have been less, perhaps even a single season . . .'."[5] Pearce points to many such historical climate events. Though not all of them happen so rapidly, many happen rapidly enough to indicate that dramatic climate change with serious impacts has been a periodic part of Earth's functioning. There is no reason to believe that anthropogenic climate change is exempt.

Intensified singular weather occurrences like superstorms have become one way for people to connect to climate change. Pearce discusses jungle fires, hurricanes, and droughts that in 2007 already seemed extreme but which have been super-seded by other events since then. Much discussion has been given to the question of whether climate change caused these weather events. "The question is not: Can we prove that events like Mitch are caused by climate change? It is: Can we afford to take the chance that they are?"[6] Pearce encourages proceeding with caution, even if scientific certainty about the role of climate change cannot be established.

Yet climate models suggest probable large-scale climate events that dwarf such extreme weather events. For one thing, the Greenland ice sheet, which was once thought to be incredibly stable, turns out to be subject to a number of feedback that produces rapid and dramatic change. The collapse of the Greenland ice sheet, a process that was thought to take centuries, is now "very likely" and could hap-pen within a few decades raising sea levels 23 feet and potentially setting off other irreversible climate feedback.[7] Alternatively, there are a number of different aspects of climate change which could interfere with the Asian monsoon. Over three billion people rely on the monsoon rains for food and when it has faltered for even one year in the past, it has resulted in the death of tens of millions.[8] Any of the potential threats to the monsoon from climate change would likely cause it to be regularly or permanently disrupted, resulting in widespread suffer-ing and starvation throughout Asia. Events such as these, immense sea level rise, extreme temperature increases, mega droughts, and other climate disasters may be in store. The point is not to produce an apocalyptic mentality, but rather to use past and current events to orient us toward potential futures and thereby influence our actions today.

There are a number of mechanisms that drive these rapid and uncertain events. The scientists that Pearce discusses use the language of feedback, thresholds, amplifiers, chaos theory, fractals, phase space, complexity theory, and bifurca-tions to explain climate change. What unites these terms is the vision of climate that they describe. Climate change cannot be seen simply as increasing tempera-tures resulting from increased carbon emissions. The climate is actually a system composed of a number of interconnected elements such as stratospheric winds, the ocean conveyer, El Niño, the biological pump, solar pulses, the hydrological cycle, and the Earth's wobble. A change in one element may set off changes and feedback in others. For example, in the ice-albedo feedback, the problem is not just that ice is melting, but that as it melts it turns from a reflective to a darker sur-face, absorbing more of the sun's heat. "'You cannot at the end of the day change one bit without changing the other. They are all part of the same pattern. . . .' Each functions as an integrated system, not as a series of discrete levers."[9] This com-plex interaction constitutes a large part of what makes climate change uncertain.

Things will only become more uncertain as we learn more and new situations emerge. "Right now the only such prognosis is uncertainty. The Earth system seems chaotic, with the potential to head off in many different directions. If there is order, we don't yet know where it lies . . . the story of abrupt climate change will become more complicated before it is finished."[10] This uncertainty is magnified by the fact that our situation is "genuinely new."[11] As Paul Edwards points out, we will probably never know more about climate change than we do now because of the speed with which we a changing the climate and the lack of historical record for the direction of change.[12] Scientists have been able to trace patterns for many of the elements that constitute Earth's climate system. But human aerosol production, deforestation, and carbon emissions are new elements for which there is no record. This magnifies the intensity of the uncertainty. While the climate has many directions in which to go, humans do not. The question is what can be done so that the climate will take a direction that remains hospitable to human life.

Though Pearce's approach to climate is complex, his approach to the human response is traditional and straightforward: politics and economics will alter human behavior to some extent while technological innovation will do the most work to mediate climate change.[13] Despite his knowledge of climate denialism and his engagement with uncertainty and complexity, he falls back into treating climate change as a set of facts that rational people will use to behave accordingly. This blunts the potential of his engagement with caution, which could productively reshape the dominant modern notion of rationality, which can play a role in denialism, as Chapter 1 and this chapter show. As Marshall argues, it is necessary to move beyond a restrictive scientific engagement to build a broader orientation to this new world condition. An examination of the political potentials of the climate event moves us in this direction.

Political theologies of climate change

It is not enough to know the facts about climate change. It is too easy for that knowledge to remain abstract, to simply serve as a matter for discussion, or to remain isolated and unconnected to the movements of life and politics. The problem of those on the right who deny that climate change exists is well known. What is given less attention is the fact that most of those who know that it is occurring live in much the same way as those who deny that it exists. A scientific understanding falls short in motivating changes in the way people live. Bruno Latour calls this "climato-quietism," referencing the contemplative theology that remains detached from action, leaving that to God.[14] At the same time, other theological approaches have been put forward, which contain both problems and promise. I will begin by looking at existing theological responses that obfuscate and avoid the problem of climate change. Then I will turn to Carl Schmitt's political theology to help elucidate other responses. Finally, I will argue that Kierkegaard suggests the most productive political-theological response to climate change.

Broadly speaking, there are three dimensions to the dominant theological response to climate change. First, there is a position that downplays scientific arguments about climate change and evolution with biblical ones. This is tied to studies funded by the fossil fuel industry that attempt to discredit climate change.[15] The result is a political assemblage that fosters doubt and confusion. This goes beyond the scientific uncertainty analyzed by Pearce because it mobilizes social and political institutions into a machine that produces doubt.[16] The inclination of doubt is to wait to see what will happen, if anything is done at all. Chapter 1 took up Kierkegaard's analysis of how doubt feeds into further countermeasures, however, such as comfort, job security, a regularized media, and legal force behind it all. These countermeasures work with doubt to ward off any event like climate change that might upset the social and political order. "They have hundreds, yes, even thousands of years between them and the earthquakes of existence; they are not afraid that such things can be repeated, for then what would the police and the newspapers say?"[17] The institutions and norms structuring society have no capacity to accommodate climate change, which calls into question its interpretation of the world and the expected trajectory of its future.

Second, political leaders who publicly profess certain theological visions use political power to pursue those visions in ways hostile to established and scientifically supported public programs, empowering those sharing similar theological views to do the same. Under Ronald Reagan and George W. Bush, "many professional environmental scientists and highly competent career civil servants were fired or forced into early retirement, replaced by others with apocalyptic religious views and considerable hostility to laws and regulations aimed to protect the environment."[18] These appointees administered the law and adjusted scientific reports according to their beliefs.[19]

A third theological response is a fundamentalist Christian vision of the end times. According to this interpretation, heaven is the true reality and Earth is just a battleground in which the forces of good are pitted against sin and human evil. The only solution to this situation is Christ's second coming and redemption. Yet this redemption is tied to Armageddon on Earth. According to this theological response, intensified and more frequent climate events are signs that the end times are coming. Thus, fundamentalists feel no motivation to care for Earth, continuing instead to promote and engage in behaviors that drive climate change.[20]

Carl Schmitt's *Political Theology* can help elucidate some of these theological responses to climate change as well as others that initially may not seem theological. His argument that "all significant concepts of the modern theory of the state are secularized theological concepts"[21] was discussed in Chapter 4. Just as God intervenes in the normal order of the world to produce a miracle, Schmitt argues that individuals, usually powerful leaders, suspend the legal order in order to carry out other extralegal measures. There is no constitutional or legal order which can anticipate and account for every problem that will arise. Sometimes these problems lead to situations where the sovereign intervenes to maintain the existing order or establish a new one.[22] Schmitt believes that there is no system so secure that it can eliminate the exception, which means that the world remains a perilous place.[23] Schmitt's personal interest and extreme politics infused this analysis, leading to a skewed view of peril in terms of general risk to the state.

Chapter 4 argued that another view is possible. Schmitt drew on Kierkegaard for this theory of paradoxical exceptions that interrupt an established system, upsetting and possibly transforming it. For Kierkegaard the exception exposes the individual to existential anxiety; for Schmitt the threat of violence. While Schmitt favored state security, an alternative response is to assume the pregnant risk of placing oneself outside the legal order in order to transform it, through civil disobedience for example. This might be closer to the discomfort that Kierkegaard urges us to work through. This more uncertain and anxious approach to the political exception was buttressed by Catherine Keller's apophatic rebuttal to Schmitt.

Someone familiar with Schmitt's worldview would not be surprised by the way that Reagan and Bush replaced environmental and science administrators with those in line with their own political and theological views. This is an instance of sovereign power obscuring the uncertainty of climate change out of commitment to a neoliberal vision of security and stability. Reading Schmitt also allows us to see how other responses to climate change are tied to this particular expression of theology. First, extralegal measures are already being deployed in anticipation of emergency situations brought on by climate change:

> Since the 2008 economic crash, security agencies have increasingly spied on political activists, especially environmental groups, on behalf of corporate interests. This activity is linked to the last decade of US defence planning, which has been increasingly concerned by the risk of civil unrest at home triggered by catastrophic events linked to climate change, energy shocks or economic crisis—or all three.
>
> Just last month, unilateral changes to US military laws formally granted the Pentagon extraordinary powers to intervene in a domestic "emergency" or "civil disturbance"[24]

Those with power refuse to address the global uncertainty of climate change to maintain security and comfort for a well-established group of people in one country through whatever emergency measures are necessary. Indeed, the US security apparatus already concluded in a 2007 official report that it is "essential that the impact of climate change is systematically built into national security and defence planning."[25] On the basis of this and looking at the destabilizing effects in other countries around the world, Welzer concludes that the connection is a direct one: "climate policy is security policy."[26]

Professor of Environmental Studies and Politics David Orr has a different but related approach to climate change. He refers to the situation that we will be in for centuries if not thousands of years as "the long emergency,"[27] which includes not just climate destabilization but also the end of cheap oil, ecological degradation, and other linked social and natural problems. He argues that during the long emergency, "government will be required to take unprecedented measures . . . the capacity of emergency management will have to be made much more robust and effective, not just for intermittent events but for multiple events, which may occur regularly."[28] While Schmitt argues that we always need to be alert to exceptions that interrupt the norm, Orr suggests that climate destabilization will be so intense that the exception will become the norm. The most important and necessary response for him comes through governance and strong leadership. Orr is aware of and worried about the authoritarian tendencies of this position and does try and specify the kinds of actions that future leadership should take, emphasizing education, being humane, and condemning military action.

Schmitt's political theology, however, helps us see the connections between a vision like Orr's and the one pursued by Reagan, Bush, and Obama's military. Though there are many differences between them, they all rely on the efficacy of state sovereign power relatively insulated from popular accountability to make emergency decisions. Orr's vision presumes the emergence of an effective and humane leader who can educate and convince the public while maintaining extensive democratic support. But the two exemplars he draws on, Lincoln and Roosevelt, both engaged in dictatorial practices to carry out their policies. Orr underestimates the difficulty of maintaining support for leadership as problems escalate, intense changes to ways of living have to be made, and results are slow. Such a situation makes militaristic control more likely. It may not be wise to promote a political response to climate change that emphasizes the necessity of sovereign power. Such responses tend to deal with the uncertainty of the event by using sovereign power to minimize and control that uncertainty, which is why Mann and Wainwright reject such Climate Leviathan solutions.[29]

Kierkegaard's thought suggests an alternative response that engages the event positively rather than allowing the lure of comfort in a secure and prosperous society blind us. Criticizing those who ward off anxiety and uncertainty, Kierkegaard pushes us to work out solutions based on these feelings. Climate change produces a condition of anxiety through the uncertainty of what will happen, the uncertainty of how to respond, and the vital threat to individuals and species. We are in the position of Abraham, who was unsure of his relation to the event of God's command and unsure of what to do with the life of his precious son on the line. This anxiety is incommunicable: each person must sound out their own relation to climate change, to their mode of being, to their expectations about life, to the future of their family and community, the kind of world they live in, and the future other species. There is no formula for how one comes to terms with this, incorporating the uncertainty, distress, and anxiety of the situation into one's life. This is one reason so many people resist or ignore the issue and why so few positive responses have emerged.

Kierkegaard's response of faith is reached through a decision in which one cannot rely on others or expect a calculable outcome.[30] For both Schmitt and Kierkegaard, the event presses the individual to a decision, but each draws on the existential in a different way. For Schmitt it is a matter of concrete life and existential threat, whereas Kierkegaard is concerned with how to deal with the internal tension of anxiety in an individual. A Schmittian decision tends toward securing social order; a Kierkegaardian one gives up the desire for the security of order to live through anxiety. Kierkegaard suggests that if we come to terms with the anxiety and uncertainty of the exception, we can live better in a world of discomfort and insecurity as well as affirmation and joy. Instead of using uncertainty to push away the event, we need to embrace it as a source of energy to help us commit to the event. This dimension is helpful for engaging the apophatic democratic dimension of politics that Keller urges against consolidations toward security.

Kierkegaard teaches us to have faith in this world. As difficult as it might be and as much as our desires might try to carry us beyond it, human experience is caught and contained within it. His aspirational character of the knight of faith "belongs entirely to the world."[31] Kierkegaard shows faith as material, rejecting otherworldly versions of belief. Deleuze draws on Kierkegaard to expand this call for faith:

> The modern fact is that we no longer believe in this world. We do not even believe in the events which happen to us, love, death, as if they only half concerned us. It is not we who make cinema; it is the world which looks to us like a bad film . . . The link between man and the world is broken. Henceforth, this link must become an object of belief: it is the impossible which can only be restored within a faith. Belief is no longer addressed to a different or transformed world . . . Whether we are Christians or atheists, in our universal schizophrenia, *we need reasons to believe in this world.*[32]

To what extent do our lives, in how we succeed and consume, in how we love and learn, in how we relate to friends and other beings, and in how we dream and commit look like a bad film, disconnected from the natural processes that sustain and support but also threaten us? If our connections to the world are anchored in the desire for regularity and constant growth and improvement, then a Kierkegaardian faith can connect us to a world replete with volatility, disruption, downturn, and periods of irregularity. The world around us is shifting and we do not notice and respond because we do not believe in this world. Instead, we prefer fantasies that support our comforts, desires, and the existential security of a routine and calculable world. More intense storms, droughts and fires, rising seas, forced migrations of people, intensified state security apparatuses, and the increasingly looming threat of crossing a tipping point that would set off runaway climate change impinge upon our lives, presenting opportunities to reforge our link the world, to begin to see, hear, and act in it again. It is necessary, then, to examine the bodily, mental, and social linkages that lead us to keep intensifying climate change and how they might be altered.

The ethics of drives and habits

Scientific information about climate change encounters further problems when it encounters emotions, psychological routines, engrained habits, and social norms. As discussed in Chapter 1, the systematic ethics of the analytic liberal justice approach have problems dealing with these elements of life as well. The retreat from culture, norms, emotions, and habits into ahistorical reason amounts to a failure to deal with significant drivers of climate change and thus misses key points of intervention for shaping more effective and ethical responses. Clive Hamilton's *Requiem for a Species* does better when it frames the problem as one of coming to terms with a catastrophic event.

> Most people do not disbelieve what the climate scientists have been saying about the calamities expected to befall us. But accepting intellectually is not

the same as accepting emotionally the possibility that the world as we know it is heading for a horrible end . . . No one is willing to say publicly what the climate science is telling us: that we can no longer prevent global warming that will this century bring about a radically transformed world that is much more hostile to the survival and flourishing of life.[33]

This alters the approach to climate change in a few significant ways. The problem is no longer the facts of the event, but rather our unreadiness to internalize them as part of a different vision of the world. It is no longer a matter of responding quickly to avert disaster, but of trying to shape the form of the disaster and ourselves with it. We no longer have to adopt a more environmentally conscious mode of living, but also adapt to a very different world than one we have ever known. It is through this approach that Hamilton seeks to explain why we have failed to act so far, why we may continue to do so, and what the implications may be.

Though the scientific information on climate change does not alone produce an adequate response to the problem, it is nonetheless an important element in crafting such a response. Hamilton summarizes our situation based on recent findings, but in a much more intense way than is done in scientific reports:

The conclusion that, even if we act promptly and resolutely, the world is on a path to reach 650 ppm is almost too frightening to accept. That level of greenhouse gases in the atmosphere will be associated with warming of about 4°C by the end of the century, well above the temperature associated with tipping points that would trigger further warming. So it seems that even with the most optimistic set of assumptions—the ending of deforestation, a halving of emissions associated with food production, global emissions peaking in 2020 and then falling by 3 per cent a year for a few decades—we have no chance of preventing emissions rising well above a number of critical tipping points that will spark uncontrollable climate change.[34]

Hamilton does not hesitate to state the conclusion that climate change will be disastrous for the human population. This statement contrasts with the way policy papers and scientific reports tend to rely on assumptions that soften the crisis.[35] The problem is that the assumptions involved do not just skew the data in a certain way. Rather, they infuse the data with an entirely different worldview. In this view, climate change is gradual, manageable, and fundamentally compatible with our way of living. Thus, an existential acceptance of climate change may be necessary to report the data in a more meaningful way. In fact, we need a view that acknowledges that either we must radically change or the world will radically change us. As Hamilton puts it, "humans cannot regulate the climate; the climate regulates us."[36]

The problem is the way climate change is imbricated with "the frailties of the human species, the perversity of our institutions and the psychological dispositions that have set us on a self-destructive path."[37] We have not been able to accept and respond to information about climate change because of the countervailing habits, beliefs, dispositions, and practices through which we live. Even though we have already gone too far to prevent disastrous climate change, we still need to engage it. "Sooner or later we must face up to the truth and try to understand why we have created the situation that now confronts us."[38] Action can help warming from being worse, it can help delay impacts, and it is key for being ready to live in the world coming our way. How we engage it matters. A nihilistic shrug or individualistic hedonism that acknowledges the truth without connecting to it fails to recognize that nature will press change upon us if we do not adopt it ourselves. Undertaking such a change requires that we examine the drives that pushed us in this direction.

One of these drives is growth fetishism. The problem is not economic since studies have shown that the costs of restricting greenhouse emissions would be tiny, perhaps only a 1-year delay in the doubling of incomes between now and 2050.[39] "The obstacle to taking resolute action is not economic growth as such but the *fixation* with economic growth, the growth fetish, the *unreasoning* obsession that arises because growth is believed to have magical powers.[40] Growth has become invested with a symbolism that identifies it with the vitality, living standards, and happiness of a society. When economists, politicians, and commentators refuse to acknowledge that we can cut emissions while still maintaining growth, they rely on assumptions beyond economic analyses. Rather than being based on neutrality, scientific detachment, and rationality, "ethical judgments always underpin economic analysis."[41]

The same holds for science. As Hamilton notes, studies such as *The Silent Spring* and *The Limits to Growth*, "while ostensibly scientific in intention and method . . . perhaps unwittingly, called on humans to reconsider their very nature."[42] Such ethico-scientific challenges bring our lived relation to climate change to the fore. Since scientific findings both contain ethical assumptions and confront opposing ones, they are not adequate for convincing people to change the way they live. Though scientific findings can destabilize growth fetishism by showing that growth and responsible emissions policies are simultaneously possible, they can also entrench it when they legitimize an economic approach to solving climate destabilization.[43] Confronting climate change means addressing and reshaping cultural drives such as growth fetishism. But it also requires reshaping the infrastructure of consumption. Though Hamilton attends to the psychological aspects of consumption, he neglects its material and existential aspects.

Beyond its sacred position in society, growth also fuels the construction of individual identities. Hamilton argues that our growth fetish imposes public imperatives that are mirrored in individual consumption drives. Consumption is driven by the desire to produce an authentic self and the correlate impossibility of completing this task, which leads to further consumption.[44] The result is a society of individuals who have too many possessions, believe that they are too materialistic, have lost connection to more important values, and yet nonetheless continue the same behavior. This drive is a potent obstacle to cultivating climate sensitivity. "If, in order to solve climate change, we are asked to change the way we consume, then we are being asked to give up our identities—to experience a sort of death."[45] If most people are locked into this dilemma in which responding to climate change would require the death of the person they imagine themselves to be, then it is not surprising that they have been unresponsive to the threat of climate change.[46]

Most people are not used to thinking about themselves through cultural drives such as growth fetishism and consumer identity. Foucault directs us to this register when he argues that developing an ethic of the self is an important and timely task: "When today we see the meaning, or rather the almost total absence of meaning, given to some nonetheless very familiar expressions which continue to permeate our discourse—like getting back to oneself, freeing oneself, being oneself, being authentic, etcetera . . . then I do not think we have anything to be proud of in our current efforts to reconstitute an ethic of the self."[47] This idea of creating an authentic self is close to Hamilton's analysis of how we consume. Foucault does not believe in an authentic self, even as he admits the difficulty of constituting an affirmative self. If there is no complete self, then one can abandon modes of life that make such promises while seeking to modify the self for a changing world. Addressing practices of caring for oneself would give an important tool to climate change advocates trying to change their own behavior as well as those of others.[48] Similarly, Nietzsche experiments with engrained habits while attending to the way internal drives like fear, desire, and hope support or undermine those habits. Such approaches make the necessary but difficult task of engaging Hamilton's obstacles to climate consciousness possible. A multifaceted approach consisting of personal and collective efforts is necessary to press a broader public response to climate change.

One critical area for transformation is individual and social consumption habits. Yet consumerism has incorporated the goal of responding to climate change. Green products feed into a particularly problematic self-identity in which ethical self-satisfaction meets questionable consumption habits. Hamilton rejects individual action entirely, arguing that it is ineffective and that collective action is necessary to respond to the climate crisis.[49] While individual action is insufficient, the reverse is also true: someone who advocates regulatory reform and votes for the best available candidate but consumes conventionally also contributes to an inadequate response. Indeed, advocating the necessity of collective action while downplaying the importance of the individual can serve as a justification for continued hyper-consumerism until regulatory reform is implemented, even as one knows that this is not likely. The fact is that the climate crisis entails an accumulation of individual consumption changes whether it comes before or after regulatory reform. There are good reasons to change individual consumption while continuing to push for collective action. The two are interlinked, since a change in consumption helps to prepare you for political change and the latter can support and extend the former.

Hamilton further argues that green consumerism transfers the responsibility from corporations and institutions to the individual. "Instead of being understood as a set of problems endemic to our economic and social structures, we are told that we each have to accept liability for our personal contribution to every problem."[50]

Tied to this is what he sees as a negative effect on democracy: green consumerism makes the debate about individual rather than institutional problems, which empowers us as consumers while disempowering us as citizens and in the end trades off real solutions for minor ones.[51] This is true. But individual activity is not zero-sum, nor is it limited to only two modes of engagement. If people are going to use energy, they should use responsibly produced energy. This is more likely to feed into ecologically conscious public advocacy than free-market individualism, while giving more resources to companies to engage in public advocacy efforts as well. Beyond this, climate responsiveness will have to be taken up by community organizations, social movements, strategically located producers, international regulatory agencies, indigenous groups, NGOs, independent scientific advisory committees, and cultural producers in sites such as theater and education.[52] Hamilton is right that climate destabilization is a structural problem in society and culture, yet he fails to give enough attention to the fact that that structure is composed of individual as well as civil, state, international, and nonhuman actors. If the cultural growth fetish is mirrored on the individual level as a self-built through consumption, then we have to attend to the many sides of the problem, building resonances between them.

Hamilton, though, thinks that green consumerism may operate at one additional level that undermines collective efforts.

> One of the striking features of the campaign to persuade us to change how we use energy is the way the various organisations stress that we do not have to give up any of our comforts . . . Indeed, the consumption of 'green' consumer goods has itself become a method of self-creation through consumption practices (albeit a sometimes far less damaging one). By shifting responsibility on to individuals and reinforcing the sacrosanct nature of consumer lifestyles, green consumerism threatens to entrench the very attitudes and behaviors that have given us global warming.[53]

The problems of creating a green self through consumption and of using the image of being green to perpetuate consumption without really changing destructive habits are serious concerns. But individual consumption does need to be engaged, not just toward less damaging products, but more significantly toward no consumption. Hamilton is correct that there are drives operating in the individual that resist climate consciousness and that consumerist forces are already trying to capture positive energy for change and redirect it to consumption. But focusing on collective action will not be sufficient to address engrained individual habits and resistances. This is why bringing Foucauldian and Nietzschean approaches to the relationship between the individual, society, and climate change might be an important point of engagement. This is where experimentation can play a role.

Though Hamilton does not explore the issue thoroughly, he does suggest the importance of experiments that connect us more deeply to the threat of climate change. He notes studies that show that humans are more responsive to instinctive reactions and visceral evaluations of risk than they are to consequentialist reasoned judgments. Though extreme weather events may be increasing the opportunities for the former, we only confront climate change through the latter mode of risk evaluation. But "as the effects of warming are delayed, a proportionate response requires us to anticipate emotions we may feel many years hence; anticipation of feelings is a weak stimulus compared to pressing anxieties we may have about job losses or higher taxes . . . we need to use our reason to stimulate our fears."[54] Yet Hamilton believes that this approach cannot be very effective. Chapter three discussed Nietzsche's approach to carrying out such a reconfiguration of drives and experiences. Nietzsche was able to remove himself from gregarious (socially normative) concerns and tendencies of the same order as jobs and taxes. At the same time, he produced imagined intensities of experience as simulacra that were then incorporated into lived experience. Though it is not carried out through reason as in Hamilton's version, Nietzsche's approach to the event enables him to make it directly felt, suggesting that a similar procedure could be carried out with practices of consumption in an era of climate change. Indeed, a combination of information, heightened sensitivity, and exploratory behaviors can turn presumed conveniences like meaty diets and air conditioning into ubiquitous and unease-inducing cultural traps.

Foucault's analysis suggests a more critical aspect to individual responsiveness: cultivating and reconfiguring the self means producing an art of living. This art of living is a matter of making actions, habits, and discourse line up with truths to which we hold.[55] His suggestion would be to develop practices that enable us to live our lives in accord with the truth of climate change. We take advocacy for collective action on climate change more seriously from those already living that advocacy. The affective and discursive force of such a mode of living reinforces advocacy in a way that effectively destabilizes socially entrenched drives and assumptions. Renowned Environmentalist George Monbiot dramatizes this effect at the beginning of his book, *Heat*.

> Two things prompted me to write this book. The first was something that happened in May 2005, in a lecture hall in London. I had given a talk about climate change, during which I had argued that there was little chance of preventing runaway global warming unless greenhouse gases were cut by 80 per cent. The third question stumped me.
>
> 'When you get your 80 per cent cut, what will this country look like?'
>
> I hadn't thought about it. Nor could I think of a good reason *why* I hadn't thought about it. But a few rows from the front sat one of the environmentalists I admire and fear most, a man called Mayer Hillman. I admire him because he says what he believes to be true and doesn't care about the consequences. I fear him because his life is a mirror in which the rest of us see our hypocrisy.
>
> 'That's such an easy question I'll ask Mayer to answer it.'
>
> He stood up. He is 75, but looks about 50, perhaps because he goes everywhere by bicycle. He is small and thin and fit-looking, and he throws his chest

out and holds his arms to his sides when he speaks, as if standing to attention. He was smiling. I could see he was going to say something outrageous.

'A very poor third-world country.'[56]

Mayer Hillman's advocacy is supported by his mode of life. Monbiot captures his felt presence and how his intensity impinges on others. The effect was not slight since it was one critical moment spurring Monbiot to write his book. The embodied difference between Hillman and others in the room produces an event in which the way that Hillman's practices, discourses, and mode of living express his position interrupts other people's lives. It is no small matter that he smiled as he addressed such a dire topic. Individual efforts do have an effect, particularly when they feed into other modes of social and civic advocacy.

In addition to cultural and individual drives, Hamilton outlines several types of cognitive dissonance that cause people to deny, ignore, or reinterpret evidence of climate change when they are confronted by it. According to Hamilton, the roots of climate skepticism lie in the transfer of cold-war animosity from communism to environmentalism. First, calls for environmental responsibility were painted with a similar brush that labeled them anti-growth, anti-modernity, and anti-American. The second move was to reframe environmental science as politically infected, undermining the legitimacy of its claims. Finally, an institutional structure was created to actively foster alternative scientific accounts and further promote doubt about climate change.[57] The effect was to discredit and defuse findings that challenged the American way of life.

This raises the problem of whether climate science is a neutral domain free of politics. Hamilton extends his argument that ethical values underpin economic and scientific analyses. Scientific findings on the climate suggest the need for business and market regulation at an international level, if not a more assertive attempt to curtail the cultural focus on growth, technology, and consumption. "So neo-conservatives were right to identify environmentalism, and its hold on the public imagination, as a threat to their worldview and political aspirations."[58] Indeed, Naomi Klein argues that "science is telling us all to revolt."[59] Climate change has politicized the scientific domain. The result on the social level is that "rejection of global warming ha[s] for some Americans become a means of consolidating and signaling their cultural identity, in the way that beliefs about patriotism, welfare and musical tastes do."[60] The problem is not just that climate skeptics refuse to recognize the scientific truth, but that the scientific findings already imply the need to seriously change the way of living that the skeptics support. The point can be taken further. A minority within that camp likely does not care whether the science is true or not: even if the science is true, they are more committed to perpetuating a way of life that is comfortable for them than having to change to save other populations and future generations. That position, however, has far less political traction, hence the strategy producing cognitive dissonance.

Cognitive dissonance magnifies and distorts the problem of uncertainty in climate change. "For sceptics (many of whom are engineers) the return of chaotic nature seems to harbour a special fear. They are scornful of climate models because they do not predict the future with certainty, thereby attributing the irreducible uncertainty of climate systems to the personal failings of the scientists who try to model them."[61] This shows how deeply troubling the uncertainty that Pearce locates at the heart of climate science is. Even scientifically minded engineers have problems confronting it. They and many others have developed mindsets and lifestyles in accord with the assumptions of continued technological development and a human ability to shape the environment. While they have been able to repress challenges to this vision in other places, it seems more difficult to do so with climate change, intensifying the need to transfer uncertainty from the climate itself to those investigating it.

Finally, Hamilton lists a number of other coping strategies to continue putting off coming to terms with the climate threat. These include reinterpreting the threat, pleasure seeking, and blame shifting. Each of these serves to allow already established beliefs and behaviors to continue undisturbed while minimizing the significance of climate destabilization. One particularly interesting strategy is "the mendacity of hope." "The evidence that large-scale climate change is unavoidable has now become so strong that healthy illusion is becoming unhealthy delusion. Hoping that a major disruption to the Earth's climate can be avoided is a delusion. Optimism sustained against the facts, including unfounded beliefs in the power of consumer action or in technological rescue, risks turning hopes into fantasies."[62] The theme of hope promoted by many who urge climate responsiveness may be clouding our ability to come to terms with the climate crisis. The urgency of the climate problem has become such that hope, rather than giving energy to our efforts, is fooling us into believing that the problem is not as bad as it is. While Chapter 2 discussed affirmative responses that refuse all hope as well as ways to hope that do not fall into this kind of denialism, the generic version that Hamilton critiques is the most common outlook of climate campaigners.

If anyone really understands climate destabilization, it should be the scientists studying it. Hamilton went to a conference on extreme climate change to study their dispositions. The conference participants were replete with pessimism, depression, despair, dark humor, and even a celebration of non-futurism. "One, a woman in her early thirties, told the conference that she was feeling smug: 'I don't have any children and many of my friends don't want to have children.'"[63] The combination of nihilism and hedonism embodied in this response might be tempting. If it's already too late, perhaps we should just enjoy without projecting any decent future, accepting extreme climate disasters as an inevitability for others to endure. Climate scientist James Lovelock has adopted this position and suggests that people "Enjoy life while [they] can. Because if [they're] lucky it's going to be 20 years before it hits the fan."[64] This is a form of coming-to-terms with the event of climate change that has the effect of helping control population even as it undermines other forms of responsible behavior. But Lovelock sees this culling of the population as part of a positive movement according to which "eventually we'll have a human on the planet that really does understand it and can live with it properly."[65] Such a view expresses a misguided faith that humans necessarily have a place on Earth and that human perfection is possible. Thus, despite Lovelock's characteristic callousness toward human life, he is unable to affirm the contingency of human life in the way that Nietzsche and Foucault do. But his view also shares an unfortunate assumption with his free-market principles. Evolutionary and economic theories that emphasize elimination as the only method of selection fail to understand the manifold ways in which each unfolds. Additionally, they attribute to life and the market a teleology that neither has.

This combination of nihilism and hedonism seems to express an understanding of the weight of the issue, and it challenges some of the fantasies and identifications that sustain our contemporary culture. Yet what this study of the ethos of the event reveals is that such a view fails to be attached to this world. Indeed, it expresses an active will not to be part of this world at this time. Such a disposition does not accept an eventful world but resents it. Its adherents discourage others in their efforts to change and derive positive enjoyment from the misfortune of others, just as the woman who found validation in her choice not to have kids in the projected foolishness of those who do. It seems that many scientists closest to the issue may not have much to offer us in terms of a useful disposition. As the double meaning of the title of his film *Encounters at the End of the World* indicates, Herzog documents not only the McMurdo Research Station in Antarctica but also the dispositions of the scientists there toward the future of humanity. The response of one cell biologist, in particular, is telling: "Sam Bowser likes to show doomsday science-fiction films to the researchers. Many of them express grave doubts about our long-ranging presence on this planet. Nature, they predict, will regulate us."[66] Attuned to the world through the intense and fragile environment in Antarctica, these scientists do not have much hope for our planetary prospects.

How can we respond to climate destabilization when it seems that even acknowledging its full impact tends to fill people with despair? According to Hamilton, it is necessary to work through this feeling. "Sooner or later we must respond and that means allowing ourselves to enter a phase of desolation and hopelessness, in short, to grieve. Climate disruption will require that we change not only how we live but how we conceive of ourselves; to recognise and confront a gap between our inner lives—including our habits and suppositions about how the world will evolve—and the sharply divergent reality that climate science now presents us."[67] For Kierkegaard, despair can be an important part of embracing the event. It can allow us to sound out our relation to the event and start reworking ourselves to change that relation, as Eric Holthaus did when despair led him to give up flying.[68] Indeed, all of the thinkers in this study—though Schmitt to a lesser extent—emphasize that we must be willing to endure negativity, loss, and change in order to transform our dispositions. Hamilton puts it in a language similar to Foucault's limit-experiences: one should become "an active agent in his or her own disintegration."[69] This begins to open the spiritual dimension of ethos that is critical for all of the thinkers in this study, but Kierkegaard and Foucault in particular. They both called on subjects to risk their own being in pursuit of transformation. The active engagement of personal loss is an important part of an ethos that prepares us to accept and respond to change and to love an eventful world that contains loss.

Such an ethos is critical to how we respond to events. Drawing on a number of studies, Hamilton argues that a direct confrontation with climate change can produce a retreat and entrenchment of engrained behavior. A careful approach is required to connect to a nature that is sometimes hostile. According to him, the key element seems to be whether the confrontation is sustained or fleeting. An ethos that is slowly shaped to endure a prolonged engagement with death and loss can accept new circumstances and reshape itself through reflection and experimentation. Short and disconnected reminders of mortality, threat, and uncertainty have the opposite effect.[70] The goal is to change "the very way we see and understand the world, our way of being in the world."[71] Climate destabilization has been brought on by how we live. An ethos of the event attunes us not just to the event, but to this structure and to how we understand and make meaning out of it. It draws on the event to reshape that structure, produce new ways of living and new forms of meaning. Hamilton even suggests, in a way that Kierkegaard might anticipate, that we will see a return from a scientific modern structure of experience to one connected to new forms of faith.[72]

This level of what we might call political spirituality leads Hamilton to a different Schmittian engagement than Orr. The political threat of climate change, he argues, is great. "We should remember that once the dramatic implications of the climate crisis are recognised by the powerful as a threat to themselves and their children they will, unless resisted, impose their own solutions on the rest of us, ones that will protect their interests and exacerbate unequal access to the means of survival, leaving the weak to fend for themselves . . . We must democratise survivability."[73] Much like Orr, Hamilton sees the political process as corrupted, particularly with regard to environmental issues. He starts to go beyond Orr and many others including liberal justice theorists, however, in suggesting that we cannot rely on that system to respond democratically, or even to treat us justly when it is finally forced to respond. He rejects the idea that an emergency government could form which would be able to respond to climate change justly. Instead, he suggests using our confrontation with climate change to rework ourselves existentially and politically.

Hamilton urges us to produce civil disobedience events that interrupt the regular legal order to respond more actively and critically to climate destabilization. "We all value and benefit from a law-abiding society. Yet at times like these we have a higher duty and are no longer bound to submit to the laws that protect those who continue to pollute the atmosphere in a way that threatens to destroy the habitability of the Earth. When just laws are used to protect unjust behavior our obligation to uphold the laws is diminished."[74] This is an ethos of pregnant risk that resists and even seeks to preempt the responses of a Schmittean ethos of security. It involves pressing the law at critical points, picking strategic moments to break it, and risking ourselves for the sake of changing it. As was discussed in Chapter 4, such an ethos can play a critical role in theoretical engagements such as those by Mann, Wainwright, Keller, and Latour, as well as practical struggles like Blockadia, the School Strike movement, and Extinction Rebellion.

It helps support deep commitment to a goal through the vicissitudes of a struggle that must also be flexible and experimental in its constantly renewed efforts.

A tragic event

In their manifesto *Uncivilisation*, Paul Kingsnorth and Dougald Hine argue that the failure to respond to ecological catastrophe is rooted in a narrative failure: today, we are caught in the myth of human progress and lack stories that can tell us where we are in the world and inform potential courses of action.[75] Amitav Ghosh reaches a similar conclusion when he argues that the arts broadly and fiction specifically have been unable to confront the culture of oil, the Anthropocene, and climate change.[76] It is thus important to engage the narrative level of responsiveness to climate change. Bruno Latour suggests one possibility when he argues that "[o]nly tragedy will allow us to measure up to this event."[77] Interestingly, many people writing about the dangers of climate change employ a tragic frame. This suggests that climate destabilization is so problematic, unfortunate, and traumatic that a tragic frame is required to think and talk about it. Though the tragedy of climate change is not referred to in a consistent way across the literature, there are a couple of significant versions.

One version of the tragic event is given in its simplest form by Nietzsche through the event of the human: "After Nature had taken breath awhile the star congealed and the clever animals had to die."[78] The insight that this type of tragedy seeks to express is that for all of the meaning and worth that we attach to humanity and its great accomplishments, it is nonetheless contingent and nature is indifferent to our existence. Humans are tempted to overvalue their place and perhaps even mistake the world as made for them or at least as being necessarily hospitable. In this vein, David Orr refers to W. G. Sebald to frame our contemporary carbon consumption. "Combustion is the hidden principle behind every artefact we create . . . From the earliest times, human civilization has been no more than a strange luminescence growing more intense by the hour, of which no one can say when it will begin to wane and when it will fade away. For the time being, our cities still shine through the night, and the fires still spread."[79] Orr uses Sebald to invoke an image of humans as defined by their energy-driven achievements, yet these achievements cannot existentially justify either themselves or their creators. This is the bind of being caught in a tragic world.

Another version is to view the human relation to climate change as a tragic one in which we bring about our own ruin: "In Ancient Greece Hubris was paired with Nemesis, the god of divine retribution, whose 'blade of vengeance . . . yields a ripe harvest of repentant wo' on those who imagined themselves to be beyond the reach of the gods or put themselves above the laws of men . . . Messing with Gaia will perhaps provide the material for the legends of the twenty-second century."[80] In this version, arrogance and an unreflective belief in the human ability to master the world lead to our downfall. Specifically, Hamilton is worried about people continuing to live carbon-intensive lives while believing that scientific progress will enable climate-engineering solutions, which themselves carry serious risks.

George Monbiot frames climate change in terms of Marlowe's *The Tragical History of Doctor Faustus*: Mephistopheles is fossil fuel, Faust's magical abilities are our fossil fueled capacities, and the flames of hell are global warming. "Our use of fossil fuels is a Faustian pact."[81] This tragic view dramatizes the tension between what we value and the danger it poses to us. Placing our climate predicament in this frame underscores the likelihood that we will not respond in time, thereby committing ourselves to a good deal of suffering. A tragic view orients responsiveness toward a greater degree of difficulty than might otherwise be anticipated.

Tragedy also raises the issue of timeliness; in this case, the "brief historical interlude between ecological constraint and ecological catastrophe."[82] In discussions of climate destabilization, it is not uncommon to hear that it is either too late or at the very least that our choice is now between a soft or hard landing. Timeliness is a critical component of tragedy.[83] As Tiresias warns Creon in Sophocles' rendition of *Antigone*: "Then beware, you're standing once again upon the razor's edge."[84] Creon doesn't believe or heed Tiresias until it is too late, but the latter's implication is that if Creon were to act wisely now, he may still be able to save that about which he truly cares. Monbiot thinks our temporal predicament is similar today with regard to climate destabilization. "Is it . . . too late? I don't believe it is. We have a short period—a very short period—in which to prevent the planet from starting to shake us off. Our aim must be to stop global average temperatures from rising to more than 2° above pre-industrial levels, which means more than 1.4° above the current point." Those words were first published in 2007, and the scientific reports have not become more hopeful. Even in 2007, Monbiot thought that there was about a 30% chance that we had "already blown it."[85] The question, then, is whether tragic possibility will be a spur to action, or a way of coming to terms with a future of self-induced suffering.

Monbiot closes his introduction and begins the rest of the book on a positive note. He switches from Marlowe's Faust to Goethe's. In this version, Faust is not dragged to hell, but is redeemed and carried off by angels because he never loses his curiosity, labors tirelessly, and eventually turns his work to improving the conditions of human life. "The gifts which threatened to destroy him are deployed instead to save him."[86] Monbiot wants to motivate the reader for the immense task laid out in the ensuing chapters. He argues that we can only have a good chance of preventing some of the most harmful and extreme effects of climate destabilization by cutting greenhouse gas emissions by 90 per cent by 2030.[87] The rest of the book goes through the details of how this might be possible while maintaining the benefits of industrial civilization.

If this is still a tragic perspective, then it is of a different sort. It is not just a matter of a universe that is indifferent to our existence, nor a critical moment for action before fate sets in. What becomes clear from this angle is that salvation is not possible. If angels carry Faust away at the end of second part of Goethe's tragedy, this only highlights that such an ending is not possible for us. Our efforts, no matter how well meaning and fervent, cannot lead to anything other than what they lead to. If we prove unable to prevent catastrophic climate change, it is perhaps likely that our scientific knowledge will progress far enough to give us some sense of the catastrophes that will befall us before they occur. We will be able to glimpse the irreversible fate we are creating. In such a situation, hope turns against itself. No matter the path we take, climate change will drive humans to bring more suffering and violence upon each other and themselves. This may come in the form of radically scaling back the comforts to which we have become deeply attached, wars fought for livable land and resources, new scales of natural disasters, or anything in between as well as violence not yet contemplated. At the peak of the Anthropocene, we must assume the burden of human finitude in a way we never have before.

The tragedy is that we are caught between two unfortunate outcomes. Creon must either undermine his sovereignty and change what he believes or lose his wife and son. Unlike Kierkegaard's Abraham, who is able to commit to the uncertainty of a spiritual report, Creon rejects Tiresias's warnings, retreats into Schmittean security, and pays the price. We must either change our values and way of life or face the consequences of an increasingly violent climate. The tragic view, however, is not focused on rationally weighing two consequences and choosing the lesser one. The point is that one cannot weigh them because one is already committed in mind and action to a certain path that appears necessary and the specific consequences remain uncertain. A tragic view uses wisdom to heed signs about an uncertain future in order to bring about painful and difficult change. As Monbiot emphasizes: "the campaign against climate change is an odd one . . . it is a campaign not for abundance but for austerity. It is a campaign not for more freedom but for less. Strangest of all, it is a campaign not just against other people but also against ourselves."[88] Tragic responsiveness does not strike us as reasonable, but as strange; it involves undermining the dispositions that make up who we are. This is why Monbiot frames the necessary response to climate change as a campaign against ourselves. Later, we may gain a different sense of abundant living. But at this point, many will likely experience the necessary change as austerity and loss. This, again, is that spiritual register of ethos that ventures our being in the unfolding of life.

Similarly, David Orr points to how "the enemy is us."[89] The idea of being against oneself in the sense of having to overcome part of oneself, destroy part of oneself, or modify oneself is emphasized by each of the four thinkers examined in the previous chapters. They all point to this kind of loss at two levels. At the abstract level, the event brings about change that involves loss: it opens possibilities at the same time that it closes others. At the personal level, one must have come to terms with this fact to be able to respond to the event. If we are not ready to endure loss, then we are less likely to be willing to recognize and engage the event. Tragedy spiritualizes loss into a disposition that makes the world more meaningful, beautiful, and worthwhile because of the loss it contains, rather than in spite of it. This is how a tragic view connects personal experience to the event and engages it in a way that seeks out its possibilities without being paralyzed by the necessity of giving things up. Austerity and loss of freedom are not ends to be avoided, but transitional states. They do not produce a vacuum, but difference. Kierkegaard and Nietzsche find joy in finitude on the other side. Tragedy imbues loss with the energy that drives Foucault and Nietzsche as they engage in personal and social experiments. It can do the same for us too.

Monbiot sees the avoidance of discomfort as the heart of an insidious denialism that takes hold of those who try to respond. "The thought that worries me most is this. As people in the rich countries—even the professional classes—begin to wake up to what the science is saying . . . our response will be to demand that the government acts, while hoping that it doesn't. We will wish our governments to pretend to act. We get the moral satisfaction of saying what we know to be right, without the discomfort of doing it."[90] This is a different and more complex kind of denial: the only thing that is denied, but more strongly than ever, is the necessity of undergoing discomfort. It amounts to a refusal to give anything up but is reinforced and made noble by both the moral uprightness of believing in the right thing and the displacement of responsibility from citizens to the government. This, to a large degree, accurately describes the celebration of the Paris Climate Agreement.

A sense of tragic possibility joined to belief in this world can help us confront discomforting situations with wisdom. The question then becomes not just how to endure them but how to explore them and engage the transformation they bring about productively.[91] If Oedipus had been more attentive to the signs, or more willing to listen to Tiresias, he perhaps could have given up his homeland and kingdom in Thebes and spared himself the painful truth, the loss of his eyes, and a future of beggarly wandering. Yet even the tragic outcome was not the end of the story. Instead, he lived through the tragedy and in so doing gained the wisdom to recognize the tragic world around him. Monbiot shows how we already have tragic tales from which we can learn so as to not have to endure them ourselves. Indeed, the discomfort of his proposal that we stop flying and adopt a system of carbon rationing seems minor in comparison to the projections of a future world driven by the violent forces of extreme climate change. As the forces of comfort, denial, habit, and hope show, this is not a utilitarian or economic decision about maximizing benefit. Instead, making this decision will require a tragic wisdom capable of confronting catastrophic events.

Wisdom is critical. Although science has and will continue to play an important role in mapping the climate event, it is insufficient for responding to it. It is the transition from a scientific understanding of climate change to one based on wisdom that a tragic view enables. "[Tragic culture's] most important feature lies in putting wisdom in place of science as the highest goal. This wisdom is not deceived by the seductive distractions of the sciences; instead, it turns its unmoved gaze on the total image of the world, and in this image, it seeks to embrace eternal suffering with sympathetic feelings of love, acknowledging that suffering to be its own."[92] When Nietzsche writes about the seductive distractions of science, he means the way that science focuses on knowledge as a solace against suffering and finitude. The problem is that at some point regular international flights by networks of researchers to catalog each measure of glacial melt become counterproductive. Complete documentation and knowledge are not yet a response, and wisdom does not require perfect knowledge but signs to motivate action.[93] A tragic vision gives us the wisdom to embrace suffering and with it the terror of our current predicament. It spiritualizes the energies of confronting suffering to transfigure ourselves and the world we live in. Nietzsche sees this as a movement that justifies a world of tragic possibility and connects this vision to *Faust*, the starting point for Monbiot's exploration: "'All that exists is just and unjust and is equally justified in both respects.' That is your world. That you call a world."[94] Both science and experience have already made it clear that we will suffer considerably under climatic forces in the future, perhaps more than we can now imagine. Yet a tragic view may one day be able to say "how much did this people have to suffer in order that it might become so beautiful!"[95]

Notes

1 Marshall, *Don't Even Think About It*, 2–3.
2 Pearce, *Speed and Violence*, xxviii.
3 Ibid., xix.
4 Ibid., 148–54.
5 Ibid., 150.
6 Ibid., 19.
7 Ibid., xix–xxi, 39–45.
8 Ibid., 194–7.
9 Ibid., 228.
10 Ibid., 237–8.
11 Ibid., 239.
12 Edwards, *A Vast Machine*, 438–9.
13 Pearce, *Speed and Violence*, 241–52. Harald Welzer argues that this is a typical problem of scientific texts on climate change: when they propose solutions, they do so with the knowledge of concerned citizens and without any of the understanding of social and cultural complexity involved in human behavior (Welzer, *Climate Wars*, 27–8).
14 Latour, "War and Peace."
15 Monbiot, *Heat*, chapter 2; Hamilton, *Requiem for a Species*, chapter 1.
16 Oreskes and Conway, *Merchants of Doubt*.
17 Kierkegaard, *Fear and Trembling*, 62–3.
18 Orr, *Down to the Wire*, 131; Kennedy, *Crimes Against Nature*.
19 Ibid.
20 Orr, *Down to the Wire*, 132–8. In a strange twist, some who believe in anthropogenic climate change have found it so intense that they use apocalyptic theology to make sense of it. See for example Skrimshire, *Future Ethics*.
21 Schmitt, *Political Theology*, 36.
22 Ibid., 6–7.
23 Ibid., 7.
24 Ahmed, "Pentagon Bracing for Public Dissent Over Climate and Energy Shocks."
25 Welzer, *Climate Wars*, 65–6.
26 Ibid., 129–30.
27 Orr, *Down to the Wire*, xii.
28 Ibid., 32–3.
29 For a discussion of this as well as Bruno Latour's diplomatic alternative, see chapter four.
30 Kierkegaard, *Fear and Trembling*, 79, 51.
31 Ibid., 39.
32 Deleuze, *Cinema 2*, 171–2.

33 Hamilton, *Requiem for a Species*, x–xi.
34 Ibid., 21–2.
35 Ibid., 23–31.
36 Ibid., 24.
37 Ibid., xii.
38 Ibid., xiii.
39 Ibid., 51.
40 Ibid., 64.
41 Ibid., 56. For a more thorough discussion of this theme, see Connolly, *Capitalism and Christianity*, chapter one.
42 Ibid., 37–8.
43 Ibid., 52–3.
44 Ibid., 70–1.
45 Ibid., 74–5.
46 For a discussion of death as a technique for engaging climate change, see Scranton, *Learning to Die in the Anthropocene*.
47 Foucault, *Hermeneutics of the Subject*, 251.
48 See Luke, "Caring for the Low-Carbon Self."
49 Hamilton, *Requiem for a Species*, 78.
50 Ibid., 79.
51 Ibid., 80.
52 This is what William E. Connolly has called the "Politics of Swarming" in *Facing the Planetary*, chapter five.
53 Hamilton, *Requiem for a Species*, 81.
54 Ibid., 119–20.
55 For a more detailed discussion of this, see Anfinson, "How to Tell the Truth About Climate Change."
56 Monbiot, *Heat*, ix.
57 Hamilton, *Requiem for a Species*, 98–106.
58 Ibid., 107.
59 Klein, "How Science Is Telling Us All to Revolt."
60 Hamilton, *Requiem for a Species*, 108.
61 Ibid., 118–19.
62 Ibid., 132.
63 Ibid., 204.
64 Aitkenhead, "James Lovelock: "enjoy life while you can: in 20 years global warming will hit the fan."
65 Ibid.
66 *Encounters at the End of the World*.
67 Hamilton, *Requiem for a Species*, 211.
68 See chapter two.
69 Hamilton, *Requiem for a Species*, 215.
70 Ibid., 216–18.
71 Ibid., 219.
72 Ibid., 220–2.
73 Ibid., 223.
74 Ibid., 225.
75 Kingsnorth and Hine, "Uncivilization."
76 Ghosh, *The Great Derangement*, chapter one.
77 Latour, *Facing Gaia*, 244.
78 Nietzsche, "On Truth and Falsity in Their Ultramoral Sense," 173.
79 Orr, *Down to the Wire*, 111. For similar points, see Pearce, *Speed and Violence*, xxviii; Hamilton, *Requiem for a Species*, 24; Monbiot, *Heat*, viii.

80 Hamilton, *Requiem for a Species*, 180, xii.
81 Monbiot, *Heat*, 1–3.
82 Ibid., xi. For a more extensive explanation of the issue of timeliness in climate change that goes beyond the human frame of reference see Klein, "Climate Change Is the Fight of Our Lives—and We Can Hardly Bear to Look at It."
83 Connolly, *Capitalism and Christianity*, 119–22.
84 Sophocles, *The Oedipus Plays of Sophocles*, 236.
85 Monbiot, *Heat*, 17.
86 Ibid., 19.
87 Ibid., xi–xii.
88 Ibid., 215.
89 Orr, *Down to the Wire*, 7.
90 Ibid., 41.
91 For one exploration of how to do this, see Tsing, et al., *Arts of Living on a Damaged Planet*.
92 Nietzsche, *The Birth of Tragedy*, 87–8.
93 For more on the role of wisdom and political seers today see Connolly, *A World of Becoming*, chapter 6.
94 Nietzsche, *Birth of Tragedy*, 51. The first sentence is Nietzsche's own. The second he quotes from Goethe's *Faust*.
95 Ibid., 116.

Bibliography

Ahmed, Nafeez. "Pentagon Bracing for Public Dissent Over Climate and Energy Shocks." *The Guardian*, June 14, 2013. www.guardian.co.uk/environment/earth-insight/2013/jun/14/climate-change-energy-shocks-nsa-prism.
Aitkenhead, Decca. "James Lovelock: 'Enjoy Life While You Can: In 20 Years Global Warming Will Hit the Fan'." *The Guardian*, March 1, 2008. www.theguardian.com/theguardian/2008/mar/01/scienceofclimatechange.climatechange.
Anfinson, Kellan. "How to Tell the Truth About Climate Change." *Environmental Politics* 27, no. 2 (2017): 209–27. https://doi.org/10.1080/09644016.2017.1413723.
Connolly, William E. *Capitalism and Christianity, American Style*. Durham: Duke University Press, 2008.
Connolly, William E. *Facing the Planetary: Entangled Humanism and the Politics of Swarming*. Durham: Duke University Press, 2017.
Connolly, William E. *A World of Becoming*. Durham: Duke University Press, 2011.
Deleuze, Gilles. *Cinema 2: The Time Image*. Translated by Hugh Tomlinson and Robert Galeta. Minneapolis: University of Minnesota Press, 1989.
Edwards, Paul. *A Vast Machine: Computer Models, Climate Data, and the Politics of Global Warming*. Cambridge: The MIT Press, 2010.
Encounters at the End of the World, DVD. Directed by Werner Herzog. Hollywood, CA: Image Entertainment, 2008.
Foucault, Michel. *The Hermeneutics of the Subject*. Edited by Frédéric Gros. Translated by Graham Burchell. New York: Picador, 2005.
Ghosh, Amitav. *The Great Derangement: Climate Change and the Unthinkable*. Chicago: University of Chicago Press, 2016.
Hamilton, Clive. *Requiem for a Species*. New York: Earthscan, 2010.
Kennedy, Robert F. *Crimes Against Nature*. http://digitalcommons.pace.edu/lawfaculty/582/.

Kierkegaard, Søren. *Fear and Trembling*. Edited and Translated by Howard V. Hong and Edna H. Hong. Princeton, NJ: Princeton University Press, 1983.

Kingsnorth, Paul, and Dougald Hine. "Uncivilization: The Dark Mountain Manifesto." http://dark-mountain.net/about/manifesto/.

Klein, Naomi. "Climate Change Is the Fight of Our Lives—and We Can Hardly Bear to Look at It." *The Guardian*, April 23, 2013. www.theguardian.com/commentisfree/2014/apr/23/climate-change-fight-of-our-lives-naomi-klein.

Klein, Naomi. "Naomi Klein: How Science Is Telling Us All to Revolt." *The New Statesman*, October 29, 2013. www.newstatesman.com/2013/10/science-says-revolt.

Latour, Bruno. *Facing Gaia: Eight Lectures on the New Climatic Regime*. Translated by Catherine Porter. Medford, MA: Polity Press, 2017.

Latour, Bruno. "War and Peace in the Age of Ecological Conflicts." In *Peter Wall Institute Lecture*. Vancouver, September 23, 2013. www.bruno-latour.fr/node/527.

Luke, Timothy W. "Caring for the Low-Carbon Self: The Government of Self and Others in the World as a Greenhouse Gas." In *Toward a Cultural Politics of Climate Change: Devices, Desires and Dissent*, edited by H. Bulkeley, M. Paterson, and J. Stripple. Cambridge: Cambridge University Press, 2016.

Marshall, George. *Don't Even Think About It: Why Our Brains Are Wired to Ignore Climate Change*. New York: Bloomsbury, 2015.

Monbiot, George. *Heat: How to Stop the Planet from Burning*. Cambridge: South End Press, 2009.

Nietzsche, Friedrich. *The Birth of Tragedy and Other Writings*. Translated by Ronald Spiers. Cambridge: Cambridge University Press, 1999.

Nietzsche, Friedrich. "On Truth and Falsity in Their Ultramoral Sense." In *Early Greek Philosophy and Other Essays*. Translated by Maximilian A. Mügge. Edinburgh: T. N. Foulis, 1911.

Oreskes, Naomi, and Erik M. Conway. *Merchants of Doubt: How a Handful of Scientists Obscured the Truth on Issues from Tobacco Smoke to Global Warming*. New York: Bloomsbury, 2010.

Orr, David W. *Down to the Wire: Confronting Climate Collapse*. New York: Oxford University Press, 2009.

Pearce, Fred. *With Speed and Violence: Why Scientists Fear Tipping Points in Climate Change*. Boston: Beacon, 2007.

Schmitt, Carl. *Political Theology*. Translated by George Schwab. Chicago: University of Chicago Press, 1985.

Scranton, Roy. *Learning to Die in the Anthropocene: Reflections on the End of a Civilization*. San Francisco: City Lights Books, 2015.

Skrimshire, Stefan. *Future Ethics: Climate Change and Apocalyptic Imagination*. New York: Continuum, 2010.

Sophocles. *The Oedipus Plays of Sophocles*. Translated by Paul Roche. New York: Plume, 2004.

Tsing, Anna Lowenhaupt, Nils Bubandt, Elaine Gan, and Heather Anne Swanson, eds. *Arts of Living on a Damaged Planet: Ghosts and Monsters of the Anthropocene*. Minneapolis: University of Minnesota Press, 2017.

Welzer, Harald. *Climate Wars: Why People Will Be Killed in the Twenty-First Century*. Translated by Patrick Camiller. Malden, MA: Polity Press, 2012.

Index

Printed in the United States
By Bookmasters